The Joy of π = 3.

3RD DIGIT

31ST DIGIT

314TH DIGIT

3141ST DIGIT

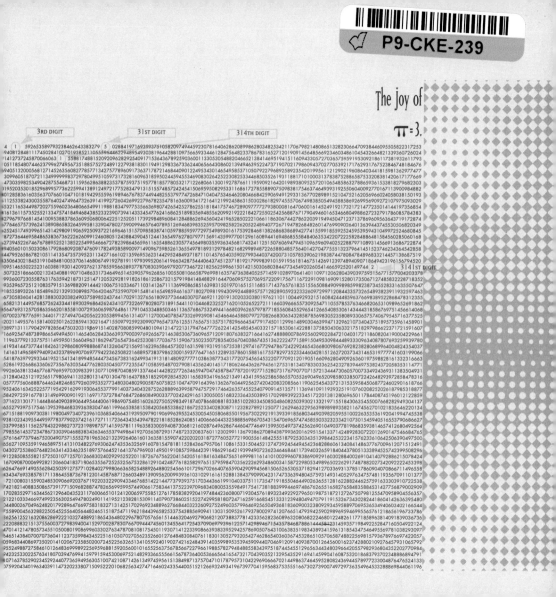

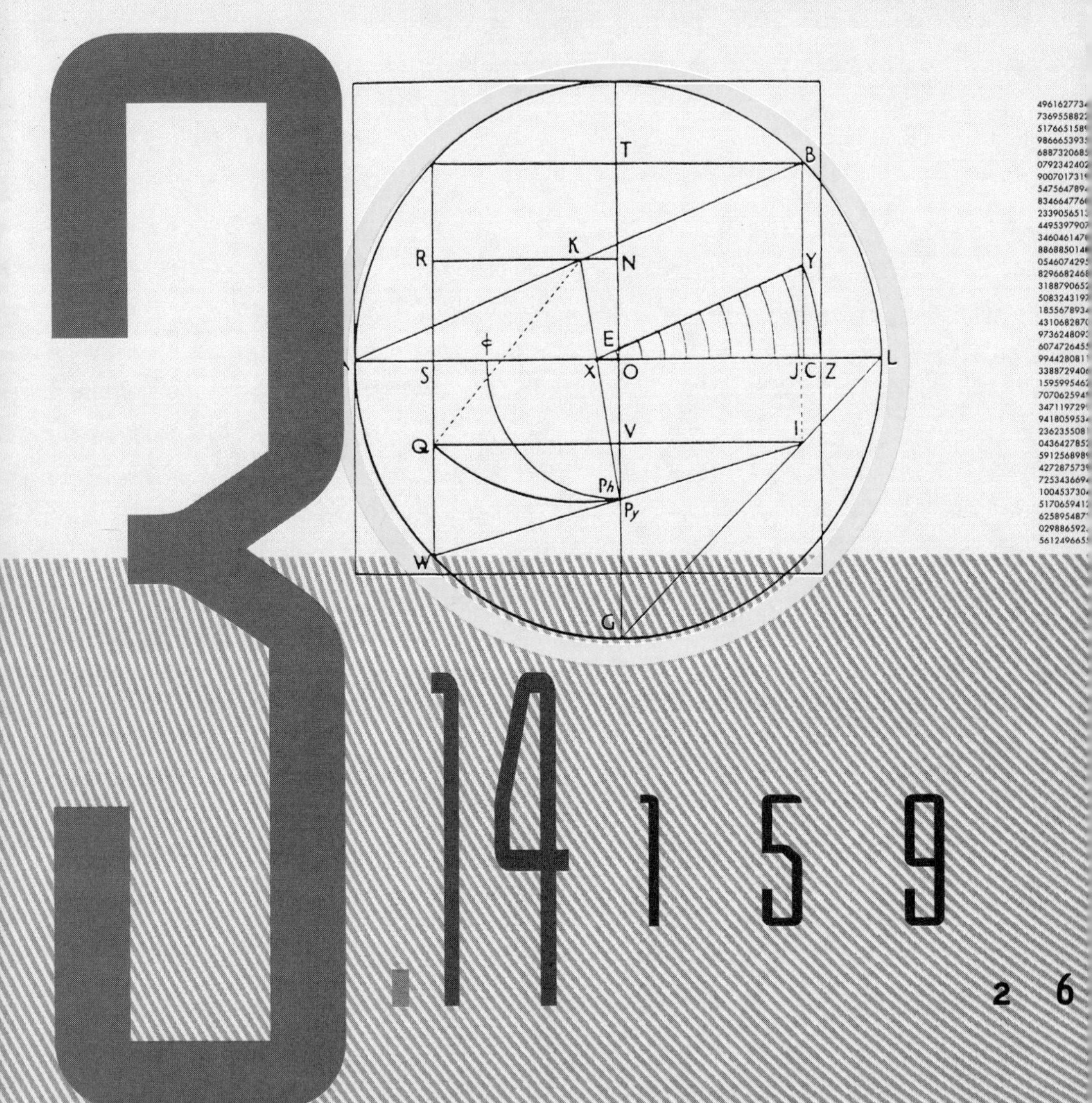

T
B
R
K
N
Y
E
S
X
O
J
C
Z
L
V
I
Q
Ph
Py
W
G
3.14159
26

496162773·
736955882·
517665158·
986665393·
688732068·
079234240·
900701731·
547564789·
834664776·
233905651·
449539790·
346046147·
886885014·
054607429·
829668246·
318870065·
508324319·
185567893·
431068287·
973624809·
607472645·
994428081·
338972940·
159595462·
707062594·
347119729·
941805953·
236235508·
043642785·
591256898·
427287573·
725343669·
100451730·
517065941·
625895487·
029886592·
561249665·

DAVID BLATNER

THE JOY OF π

Walker and Company

New York

Copyright © 1997 by David Blatner

All rights reserved. No part of this book may be reproduced or transmitted in any form or by any means, electronic or mechanical, including photocopying, recording, or by any information storage and retrieval system, without permission in writing from the Publisher.

Dictionary definitions throughout are from the Random House Dictionary of the English Language, Unabridged Edition, 1966. Illustration of Liu Hui's method of calculating pi on pages 25 and 47, used by permission of Cambridge University Press; Great Mosque, Samsarra, Iraq, on page 28, from Iraq Mission to the United Nations; cartoon on page 54 copyright Roz Chast; DILBERT on page 66 and DRABBLE on page 83 © United Feature Syndicate. Reprinted by permission; cartoon on page 69 used by permission of John Grimes; cartoon on page 76 reprinted from the "Fish and Richardson speaks your language" national advertising campaign with permission from the intellectual property and technology law firm of Fish & Richardson P.C.; "Circle Digits: A Self-Referential Story" on page 120, used courtesy of Michael Keith.

Book design by Maura Fadden Rosenthal / Mspace

Printed in the United States of America

ISBN 978-0-8027-7562-7

To Debra Carlson, who helps me see the poetry

1 9 4 8 3 3 2 8 3 3 4 7 0 0 7 5 1 6 1 9 7 9 0 0 8 6 8 7 2 8 5 4 3 9 9 6 2 9 8 1 5 7 5 6 0 5 8 9 1 6 3 7 6 2 4 7 2 3 0 6 9 1 6 2 8 7 1 1 1 1 1 3 7 6 7 6 0 8 6 4 8 0 3 2 3 7 5 2 4 5 9 6 6 4 9
3041175394613646433780467116505550467067183622128579504806716563042762671142999911348769844705037063790018109688629721757951732433802780617470496302042492916619171886243355599
2820932439194457118863215563201616542470553759386966246563341215410140322869909301591328858088312412428828763738727428380385907102927486333515030904453280525977956589205545624 3
4297982794134891756382400771612173324736428540160610044337641457220785921715591401037832020132133830963807789040957238105588293927963743816606868351950592770195153616017221589
0428785678482068291944169871819286273082704441630396254713053284388337913374768735826122116258360272896162455904189677024745382758396652299371235163048983301242141745578859159 4
2560597924277218199085562798486056174536844789237969077559455515464685316302446232567403489585456225674485820204245739199425309426422450420268903815015268360241255980759736 4
8162809304891274615119623154611400822056396786585354076686882275426503812259991620760170895567474465242344520176616503259466591296678632462137991922296145867142248249288064 76
8032108647799410041006033906792752373625460277429600734788038356687522003482457694908456862696057715701919174892260635208129737879744383548328613693956245039297680578323402171
6765559177668403757234844094617629312884926899368713898388222710602790379900190455833600797392774109266557392331470259092338905438842235132411538801855923495613993022391964505
0450369352927011566305153351918641864823442499919272027295345959906304872360804159576002966812111683172366038110542803591445720248256456105714055462420821343520948108417158289 5
7244507206354681600230512014084805435874252617101768185388355755871741542477544977222141926131555226910917556333193232224321852542218272914915981058368970250352281300214119248 6
0142480680797536996477719394906804683552808347327610306049409733091690316783097934636611832784531868716462680738833656704566010423768505801395074436479639222841126979451347730 0
4924987864965636794909929132712528977651918175427962806084932375520815361113240339713165504391887960198382138585000773242461778849187581459642642337889793330819488160040113126 5
2563569324465939840063689031525472292399141447437706963389357619260391892479363178008310261141954854360515778716004955788656579706658855104288246636305720777890226677042512681
5719795332251076389036819762844028610258805392339329474672024088541276492386447602161162620824212991660362299184923782236300983478119522913821847326342285759120979805478285250 5
9183798336801787411242644746002256241498069140074097972102327853957561512834580616541111792671042799057939449713494632895045651286884784187175802050458328387485313736911351025 5
0620102775345809439105001021833973245650472889476879298925945019875076712236379187586472012149660611512804870964886305622844083936944387216921208492008515583812510707419551872 0
8093746942459731172811721051928903896370394235776862127668210931827636649840421249381440979598631142254364839654999834790843070217643855543512574368282281530322223808347679511 1
3557014806318200453220723794891863572149106242526993994671015366846234105153338142684770627585203524099207972086991453730109551641503317628200196916411546026820723669255275141 8
4299699205398534330730680573723805041671972211273740507892726634063885068673445856077326664838457802771891147580132310551987841336521851907146068138986886710314759826461129379 5
4395266728672759948335902597445878687684964626834844341413591771458776608807784535718393293719373932364083563375766884682111179935055410208556188490102016005056395416874510822
0603555410817666460524124966224422804545243216032036019464135609792001959024049792923673298924553990101980112140290868699920575891777188074146122205024728585715367530747814389 7
3057178726836636015761361007722863196388526462351255380773194595635679653823624999265518043307963596211067455285214290262949826567553352731004687886573104724664933265679273313 4
5122955059186232937393326086077451350775309015744438294873397796053228493583013618379586264803212973684748175164769136621103603695091066665051717115082782009327883587225983940 4
6306837631811808904423626219988123682680785795262197216687201745517472627818032683058548803970977047934831035439855907843552776676033139884605271503138856332467688927104595851 9
3289513916782385773577265810047982563935519352005520408002870596782497393747886052835649359149783803779649600052124458347790017560424658666519980770288394385163809550430492196 0
3244360903400851746604296274309768387151945982644735940234248211044757291117795877313413553609527595708986125867714562523994500759380206093550248920084767332293085742222550206 4
5569023912654366357852427242905605320575403082101451238209021746697579765347517250146583747884808053773515042222404295760361375432486199655891939220504699982106293160967565179 0
7513229607778575533102658584257608668676453552092774827556754517716995087894118059363052499449670123759800655349987396663953944170170596981015127193331184076792327185395398097 6
4048527846743872316432910029065495308612833302664007580129618499207022002555972156957588376168784364346792755863573972253564884133060119289574642809357858081132331433115287482 1
7976603971257952890036407198923328131611640416937736628013259738222237426818917648959642270338039059295964969648213311447316676504197678110849096646942571706945700787126401448 6
5224284694889761725674653522050616210730010192624831468212035516995015220073163840041320303332423121670826854689317584366304307843507859281044784926639526523987186441733800856 8
1692321347429754583269402161253332837900960648627785494126679513674045877416945596140762656625029900692267267876036587137932796041848839393399346926354341548095183623323317522 93
7035210291464133127520371171667548720634738923293785107290295144629274154676194794274716691603049782928896147458702649979707920638724082502300642554499590401197410853516784440 9
0188046293748354439614400353523310304041178457228902958180581032123743825898702747370401068377771592512645357065083009214792583498924751274536220061058545759973693135297078143
7428413405519544672148941505745283917160371545308252555834320251254241662445752456296445791076971715214709518505500355054390631688258105785074635656204791466768055698438455202
7709969719889807233714869563567031776877637897432734928293439051455670607446079704769316462781214171381827437856146219708807021062411057377851471358837377388240765280451914271
3748811055974471831009393751976598021002410125112308136826033847449108771613228576602639388492849598982365657270426357202637482564949491262914191713064628059566982549360326132
0192528043461704390289260279931404361370265820121312851488158573111782104131033572888718172952627112000814750640268304641898876974787917317370381399918882424169942121527760451 8
5956711909418073734793310997092831554681656395271010461137625406644958618385463898220899677832955011143149959368039822230371363295742321735744647342109741491743641994731958840 0
5263872695923183642325491845595504534377846709470450595420120211422086419127904935994521373924871107432314951138042937936554363721726348190757113531270930795272952211247953149 8
9699080894665747695565124360561142008663990560990003803025061242360775032934134728905013167728097131626834959634092922430311950848788671035335200237127302029165929752526570392 1
0421496349523857085605723434621576956985134068304548331545907536471146996824209102321431171769227738534770417794076441001301048596092707211320523185382227444870243327103987811 4
7912754608083611568779215131131045008366363100751751102590028086427715020962713662397401075288445468331618211502789264307297635576105511246203324800531059951115054314848295534 3
2959830574272451737886527193000732321736237587327314890910945537402704811855571990516839387453520679708592118964078548950410940569965988715988633620779550452193215633612468530 3
1747054439402941829263552401554523160986825531389701880153970459625016917966481250155592311482673005633835797260328601778474149600456972578349562058732873012451455576345230298
6481495441009078835298012070126541095251846066620176742045257367994690771908453787482060802904825167017661982073061833123921935356900407052154989390344659388090477241695436 5
1858075066490459443188862978723571603022481352204601096352145082806397492755128476943549962033991644887919743790209571888632002475020791023790730729637463263366745942755637845
3569136734552401489712590948036856628232100500394007310663207525728314711519263328928520696723934717509829526021254947643301953574383509258283111339115390633766173730772363027 9
8898699857994501659237690675488379889294006051628261400481504694828140330839164342486509396354589091328059511163345503656348245191505831794980831827281347950507727173359496633 7
1882149192837871164639035669257799424573943554730449355593968480327902086141968150826064810924688543383329866390745478052636291615627988031878282707451630327863907566533621975 0
6322424864576945975356927980063898262930000761251494798008956712452569559827585485769012463686594942242277271775184964175107159841635720724122437196806720392706478942789421 71
2842641334271183184794413346064724314115015509855117124146682433123520628406572269260690474791964472975283227495698196327787281625954012020538073295825004974459308097824095299 1
2965423318498798800771681631986086512088315867256506594414061844683749631892913745993421603484822883158289730942161473689255851699271553115588887600721703410245587440208443428 2
7300467309795555666811501300338889583802316464313829002600763228503475830780788958180313981020762788985174353478225120846759497430024437895842895680752663203627696299460180834 94
1994912706559130840005862656399639110406851041282007153246256426371456355757694528492711263557719632506589654553648212545926335525729259528149934158787765156922311915102337344 0
7169916564763982000896984629843997759385398112133218103281989699457926176493582974837338775235285946403513823823062694536345810031936725020698280738433341175283157314342639896 4
1634712705303477569915580031181591809113788026883854757697292339888286032302997704306662886955301210272705763359897689410249968479498168420119925613480756440406559462383708723
6888125489491487948734808614168105521140018455170084444842948475507327366428272220633658240174549880829130188391401568090500008495465737300032747797209917507461785951579953202 2
3728523592040074251522563861667562031883981176186119602216284743190797025036745928280467817853664739356003540382782818456769478233745711382212193261672950104270694095202650280 5
2289859093500239449087456262053452217311940957783019536051850385496140621825306182036518273370621119899390244889753863581809944918157848783365288654365422483020278924170496896 51
1041727594750178122678581439174864942435730090917264877160595920974451146295542231002200851205225897647781148270394267766642782746259395117438071986187222655865040300284691 46
9278646800318360346381726405702707422620342971875558 09 9 3 8 6 8 7 1 2 4 0 4 6 5 6 2 2 3 3 8 9 1 4646583055430131550952851097263005080518826 5 2 7 2 6 8 5 3 3 5 3 7 2 9 3 7 3 3 8 5 6 918269371716
7730316118647494810424215127915910146065697953331337740959367493264414637024275245393350301309928336485407069840343991212452492755802997988 2 4 0 9 2 0 6 6 4 4 6 4042585966200888741 9
1649877302754037292042158109378147131362262886669454742124495528490914921933719362340294337125575556998865296623645035351920267776379424820828605689362315215237885014521313214
9146986854835944706865850109813142058926764161151621094053567807368100897342458729327052108535726763805642288409296658844777952795467107351932954747130150792208403282322044289 4
4678218396547110902117340725139724757357008555312743219996751259582568063238808838843662032622661914149347404364980000247398332092411838667429609269460701418388178110714282439 6
5779638843986478231371542498947258304114514952687242361899676305881682084632743744121039055276521871073556452571336011455804558568455865043285991767651961932711434986654077745 1
4500473072711714795712227572018128886446440777517460328242371338533765298981044232240467724632047951798097157602580088576897513405948054826877288477629384645496040270370508539 41
9092769937066804551719416040376351180185513657545109524703460226002074174282384948178225490636599208474903758320574467795910675566064077500934712981700581876940802799269046059 4
9872117634151914882251867043955731001793710004665729218037284879797156922788883970419825456570642890898582795862565990137596875007856985342094439959715236676735599115570900614 1
3018853956006933050826115788315979018829128777653969640675392080848582290475561905186375490594176472080908485239299663 6 5 3 7 7 7 4 6 8 7 0 9 8 5 6 8014236137076370467423618029218679
59 2 4 7 6 9 7 7 7 6 5292629290417983927505343294338447653339850122828362798515026374542796671771484197573390657287154305432157523544932053465375423820484485088463459 08 5 3 3 8 6 6 7 7 2 9
2 5 3 8 5 2 0 4 4 4 9 8 4 4 1 3 1 3 6 8 6 3 7 5 1 8 9 4 1 1 7 6 8486261360368193736351339325408068522692147430732913446762529322640845330844938647151561813941363435036481779475509763392559882786 90
3696323863303425794452922923775203287448902004053266813935475285501746453171721459950814556136469252665022711533738181759785579504198807548581133628915490090390806077541575736 1
3737559880187573075362487370012912238261134381039234372313536898891533749493786324984941764281417045280829693991724323286772564150483765773114493352155385230017811082761636303
7090205259503779092534110470570046565251977925679331410886632640592623178893126031528575871642421190333798725775874290129037593626972723431489357257241883794186276864566775868 6
9202760143980501638714352047767388090057892836338177973884573441001499664332358222579253517110594856078918240152199828522694650958763149247127952016446764740270468954543510306 9
8261799914022340728548915468068420957432075066211544876266446759863644388023258636088691875944227152142965066461613849638150279721730716592057826600278471814003420926569307030
9044570245964675764901852781393148131502036410498459690602253144748229457025270436304061114455142227669366501254252372074394018277525089414329152151705997 4 5 4 5 9 3 1 2 5 9 4 6 8 2 1
2 1 4 3 5 1 0 6 2 2 7 6 3 3 0 3 1 8 5 0 4 3 3 9 4 8 8 9 5 1 2 7 6 7 2 0 6 3 7 2 9 1 5 1 2 4 9 3 6 8 1 9 3 5 7 0 3 1 9 1 0 4 6 9 3 5 7 2 9 0 5 2 7 6 2 8 8 7 6 8 7 8 2 5 0 0 4 8 5 0 5 4 8 0 0 5 9 7 3 2 3 0 7 5 3 2 6 5 2 2 7 7 9 2 5 5 2 4 1 9 9 1 3 1 5 9 6 1 7 9 1 1 5 2 2 0 6 9 4 1 9 6 8 5 4 7 9 1 8 7 3 4 1 5 6 6 9 9 7 8 1 0 9 6 7 0 2 5 6 2 9 9
3993208164507741734905643398652199866390557093521198524390679861502144862392843873982018760228547123039494596615725875096503200712476657593813721248 0 1 1 3 4 1 5 3 5 5 0 6 1 6 7 5 4
7 2 0 3 6 9 5 7 9 1 0 5 5 9 7 4610671125417117453695430147191419937319722797169021161357262524311647228936664414262124385498136236949635712821160368544160710823177510780129830425381 4
1908922492085953646108213956481132053160737077720760559934981503424064077512331512158999246297497845474385785595227089267102479199199645043040166005621762962340149282181611520 5
0464381405120101763279790269327 12 2 2 7 0125927081630457940869593885030885857777676988057712027746185837281858599701772111603710982739324147197937663864843160008415792725306116408
5015150016520300200142743376390418788622635274702258984849469077694747613276391052599405660382382371636943555470658174827307182474182726362724046239944028444736424586444751046
9029976526749734435698570853905781915995859 9 6 0 9 6 7 5 0 6 1 2 8 3 0 9 1 0 1 9 4 7 4 8 8 6 5 6 5 0 7 5 1 2 6 1 3 9 7 1 3 6 3 2 9 2 7 6 4 1 5 8 3 4 9 1 3 0 4 2 0 8 3 0 0 9 5 0 8 5 1 1 0 0 4 1 4 0
7 4 5 5 7 4 4 3 7 8 4 9 2 7 8 9 8 5 7 6 0 7 2 6 1 0 5 7 6 9 7 4 1 8 1 9 6 3 3 6 9 6 7 9 0 7 5 5 1 8 8 3 8 3 2 2 0 1 7 3 4 4 3 7 6 4 3 9 8 0 5 3 6 8 2 9 6 2 6 8 7 3 2 8 5 1 8 9 3 9 5 3 0 8 1 5 9 7 2 1 3 8 4 0 9 9 8 7 5 3 6 5 7 7 4 6 6 3 5 4 9 3 2 5 3 1
1 3 9 3 6 2 5 5 9 7 8 9 5 4 3 0 0 0 9 1 1 9 1 4 2 6 7 4 0 7 5 3 8 5 9 2 5 4 9 6 9 0 1 5 7 9 7 3 4 1 9 1 8 3 7 1 0 4 0 1 6 9 9 9 1 7 9 0 0 9 4 5 6 7 8 3 5 9 6 2 8 5 7 3 2 2 4 4 7 1 4 7 9 0 7 3 2 0 4 5 6 9 6 4 7 1 9 7 8 6 3 1 5 4 9 0 8 6 2 8 4 1 2 3 3 3
2 5 1 7 4 8 1 2 7 8 4 8 2 8 8 0 9 8 4 8 8 7 6 1 0 2 2 1 0 0 9 7 4 2 7 8 3 4 7 5 1 6 4 6 2 7 9 0 5 5 3 9 3 8 5 1 9 6 6 8 8 8 9 5 6 9 6 5 1 0 8 7 6 0 6 2 8 7 2 9 5 7 4 5 9 0 8 8 9 2 8 0 1 7 0 2 7 0 3 8 6 7 2 0 7 4 0 1 0 6 0 2 4 5 3 8 9 4 1 5 1 9 5 4 7 3 9 3 2 8 1
4 2 4 6 6 2 2 3 1 2 6 8 9 2 3 6 2 6 5 0 2 7 2 0 5 6 4 0 2 6 4 3 0 2 1 7 7 6 9 0 3 1 8 9 5 5 5 5 2 0 6 1 1 2 7 1 1 4 6 3 1 4 6 7 1 7 0 3 8 9 1 5 7 7 3 3 9 0 0 6 5 4 5 2 8 6 9 2 3 2 7 2 0 8 0 8 1 1 1 5 7 8 7 5 7 3 7 4 9 9 1 0 3 5 3 2 4 4 4 6 6 9 3 6 1 6
5 3 5 1 7 5 2 2 1 2 4 6 8 8 6 6 0 8 0 5 9 3 9 7 3 8 0 5 4 6 8 9 4 8 6 7 5 5 6 0 2 5 8 8 7 0 6 8 7 1 0 3 0 8 1 1 8 9 8 9 2 2 0 2 4 2 1 7 4 9 5 2 9 3 4 5 8 2 1 9 5 3 5 3 0 0 9 9 1 5 6 1 3 5 5 3 6 0 7 3 1 5 9 0 9 5 6 7 3 4 6 9 9 0 6 9 9 2 4 8 7 4 2 6 8 0
0 1 9 5 3 8 2 1 7 5 2 4 6 2 1 0 5 3 4 9 8 6 2 7 0 1 0 6 1 3 2 1 5 9 0 7 5 7 2 6 0 2 4 0 8 0 4 3 0 0 8 2 7 8 6 8 3 5 6 2 9 3 1 9 8 3 8 4 2 7 1 0 5 2 1 9 8 3 5 4 7 2 7 5 1 1 7 6 4 2 3 3 0 2 7 9 9 5 8 9 2 6 8 7 2 7 3 0 5 3 1 1 8 3 5 5 8 0 5 6 8 7 5 2 7 6

ACKNOWLEDGMENTS

This book, more than most, has truly been a collaboration among many individuals. I would particularly like to thank everyone at Walker and Company, including George Gibson, Liza Miller, and Marlene Tungseth; the designer, Maura Fadden Rosenthal; Stefan McGrath of Penguin UK; my agent, Reid Boates; and my research assistant, Suzanne Carlson. Many people throughout the world contributed valuable information, including William Gosper, Stephen Wolfram, Peter Borwein, Simon Plouffe, Richard Fikes, Allan Fallow, David and Gregory Chudnovsky, Danny Bobrow, Underwood Dudley, Gene Golub, Barry Fisher, Vaughan Pratt, David Nelson, Benjamin Ginsberg, Larry Shaw, Ivars Peterson, Eve Anderson, Sandee Cohen, John Grimes, and Jon Winokur. Thanks also to Ted Nace, who told me about Richard Preston's 1992 New Yorker article on the Chudnovskys and helped catalyze this book.

98193813214238441581929070046389771683887207916361741432497079109658162746429717072871725142745898356897095534626820169085356108944898407
10058192030217694512077177458879551951047338418473998079630676788584516757572990430697154264238349800987086993367091210839445350624592243
23123482785496603746571880148929379451478705406079245759006012196221239287200172155886663457349714095337211516559857579417244198890261670
16101611557834315025460328781198424027484608510722406676778760855247617773833089502610064388350550205456324346167859451941795669874968515
24488384751361818066710831616556420936927052061189851729261714171443465550870630606355101294940030975916779915842604919712095432270267843
26542965724032720887143219996453132025871096771651285496699625526986073117637182074988273997706019913620930832307368382064557325637659829
12578131492224220427971241441629951265945639792759380383804782623160424325399132851123032247037561942321733047854078576244013291717992979

24078339071575798142681686465538294684739920588863165593491986789696284044734496802407709283137640810335225524271740410767356542444100448
33474401017264410529547872963458986405012036080244511903509949744939736171815752770937802092366681358416362683192634067141827974213425462
20705415600050959674045616840451777174795279035325493258912048338574659009678173041600005210889346107687540042419777803082885181200173369559
12713771419501136130440975327919050489158324639914348353164868154857917863293512392555251021118278857369606027693130146966143344964230211
43824837056335327938588952676720766889712744358156320881066501495681435587965769098577659027687074536592763649755534496173080781609871032
48013795136170367763457594975686208013996374551762425147780628722265971455482906769295713643572152674468987889418820751292225756509143552
828874614195097862427527881571566400763721037803194043095844272549269987169234331890022141503111399876526068876156674021019720171960239086

108297492763956954115303227546017380709569259935798530244434767163995914623179312399899869284379570249236955158729768385400522765149561444710597196288988815710941517170151811474351364385400511624620213117480079198374970001004713634325232815789113554504533719052750682291561850033284695679262262208190442473330362503892792071585960039363153688427243753660970968479347411331983286194414606539227840999031438403545650470567895520248271760118743356436902435030856313095590552503049273161331173492258464460902453507919018441129932169977045183285335864804285568222087372136164905863032563689130841037602156799270200053223354398046531193397754590440450785680213984650090693429547310269249947586466058091669984160684646087293943808274308285817479694172872990311013192675573897984091364253479694943480377703364634958476862982590103470727861218623001986607987782684245933835638919570206853521603211635230064988744600200170413056985365154668752023859375183280372851143274

THE SYMBOL π 75

Why π? A brief look at the most commonly used Greek letter, π, to symbolize the ratio of a circle's circumference to its diameter.

THE PERSONALITY OF PI 83

Numbers, like people, have qualities and characteristics—you can get to know a number and explore how it interacts with other numbers. What is the nature of pi?

THE CIRCLE SQUARERS 89

A lighthearted look at attempts to solve one of the most famous (and elusive) mathematical puzzles of all time: constructing a square with exactly the same area as a circle, using only a straightedge and compass, and doing it in a finite number of steps.

MEMORIZING PI 109

Pi mnemonics, poems, and other tricks to memorize digits of pi—including the mind-boggling stories of people who have memorized and recited tens of thousands of digits.

AFTERWORD 122

FOR MORE INFORMATION ON PI . . . 123

INDEX 124

Preface

CIRCLES AND SQUARES

Mathematics . . . would certainly have not come into existence if one had known from the beginning that there was in nature no exactly straight line, no actual circle, no absolute magnitude.

—**Friedrich Nietzsche**, *Human, All Too Human*

It seems like such a simple problem: Draw a square that covers the same area as a circle, using nothing but a straightedge and a compass. How hard could that be?

On the one hand you have a circle—the simplest form in the universe. A raindrop in a pond produces perfect circles of waves that expand indefinitely until canceled by the friction of the shore or by the perfect circles made by other raindrops. The branches of a tree, when viewed from above, form circles around their trunk in an attempt to achieve optimal surface area for soaking in the sun's rays. Even planets and stars try to form circles and spheres in space, though gravity and spinning forces push and pull their pure mathematical curves into the complex forms we see in nature.

Circles are everywhere in the natural world, and to the peoples of early civilization, the great circles of the moon and the sun looking down on them each day were sources of infinite power and mystery. Even before civilization began, people probably drew circles in the sand with a peg and a rope, building their own infinite forms. The earliest homes and sacred sites, dating back as far as 8000 B.C.E., were circular, owing perhaps to religions based on reverence for the Earth, the mother-goddess.

On the other hand you have a square—exquisitely formed with four equal sides and four equal angles. Since the earliest recorded history, the square has been the opposite, the antithesis, of the circle. Squares are found rarely in nature—perhaps only in the purest of crystalline structures. Where building a circle comes naturally, we have to measure and calculate to create a square. The simplest squares develop from circles: When you draw two perpendicular lines through the center of a circle, their ends form the corners of a square.

Squares have become symbolic of our human ability to measure, to solve, and to partition. Where circles denote the infinite, squares indicate the finite. Where circles reflect the mystery of the natural world, squares enabled early civilizations to segment the land for farming and for ownership. We no longer live in circular homes, preferring instead the defined walls and angles of our modern houses.

8728543996298157560589163762472306916287111113767608648032375245966493041175394613646433780467116505550467067183
6221285795048067165630427626711429999113487698447050370637900181096888629721757951732433802780617470496302042492
9166191718862433555992820932439194457118863215563201616542470553759386966246563341215410140322869909301591328858
0883124124288287637387274283803859071029274863335150309044532805259779565892055456243429798279413489175638240077
1612173324736428540160610044337641457220785921715591401037832020132133833096380778904095723810558829392796374381
6606868351950592770195153616017221589042878567848206829194416987181928627308270444163039625471305328438833791337
4768735826122116258360272896162455904189677024745382758396652299371235163048983301242141745578859159425605979242
7721819908556279848605617453684478923796907975594555154646853163024462325674034895845462256744858202042457391994

And so back to the problem. For four thousand years of civilization, it has seemed natural that we, as humans, would be able to find a mathematical, geometrical relationship between the circle and the square; that we could measure the common circle as we measure squares or even triangles. It has always seemed logical, paradoxically, that we could discover limits to the infinite and could somehow calculate nature. But we were wrong.

pi[1] (pī), *n.*, *pl.* **pis.** **1.** the 16th letter of the Greek alphabet (Π, π). **2.** the consonant sound represented by this letter. **3.** *Math.* **a.** the letter π, used as the symbol for the ratio of the circumference of a circle to its diameter. **b.** the ratio itself: 3.141592+. [1835–45; < Gk *pî, peî*; used in mathematics to represent Gk *periphérion* periphery]

Why Pi?

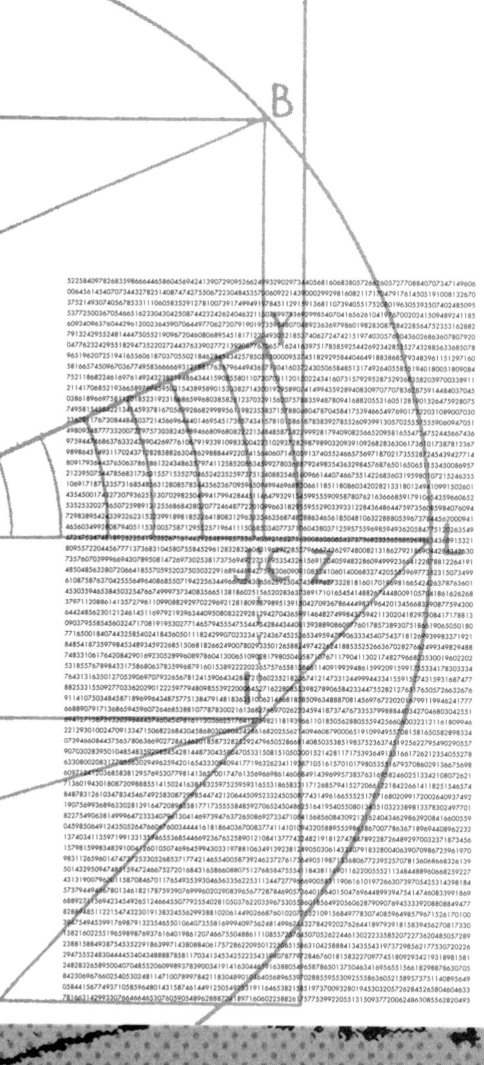

Probably no symbol in mathematics has evoked as much mystery, romanticism, misconception and human interest as the number pi (π).

—**William L. Schaaf,**
Nature and History of π

In our world of high-tech precision instruments, where we assure ourselves that perfection is attainable, it's hard to admit that we cannot truly solve a problem as simple as dividing a circle's circumference by its diameter. And yet this value, represented by the symbol π (pi), has puzzled mathematicians for nearly four thousand years, generating more interest, consuming more brainpower, and filling more wastebaskets with discarded theories than any other single number.

Of course, with just a tin can and a piece of string you can find that a circle's perimeter is just over three times its diameter. With a good metal ruler, measuring by the tenth of a millimeter, you can even see that the ratio is just over 3.1415 to one. There are methods of calculating the ratio even more accurately, letting you find 3.141592653 . . . , where each additional digit represents a value ten times more precise than the last. But as hard as you calculate, and as clever as you are at finding new techniques for measuring, you'll never find an exact value for pi.

Nonetheless, mathematicians throughout history have dedicated years of their lives to churning out as many digits as possible. The current record, over 51 billion digits, is a testament to the incredible power of both brain and computer. But why would people do this? No measurement realistically requires even 100 digits of pi. In fact, even the most obsessive engineer would never need more than 7 digits of pi, and a physicist wouldn't use more than 15 or 20. So why are these mathematicians so driven?

This is the question that partly inspired this book, though unfortunately, as with most things in life, there is no cut-and-dry answer. Certainly, calculating pi is one of the hardest workouts with which you can challenge a computer; a single wrong digit can cause every subsequent number to be wrong. In fact, whenever individuals have attempted to break the world's record, they have uncovered deep underlying flaws in their hardware or software that would have been almost impossible to identify any other way.

The quest to understand pi often has less to do with actually calculating more digits than it has to do with searching for answers that could explain why something as simple as the ratio of a circumference to a diameter should unfold in such a complex manner. The search for pi is deeply rooted in the human spirit of exploration—of both our minds and our world—and in our irrepressible drive to test our limits. As with climbing Mt. Everest, people do it because it's there.

In the story of pi we find both the mythical and the mystical, the profound and the profoundly silly. Pi teaches us about the limits of our own comprehension, clearly marking the boundary between the finite and the infinite. We know pi best from the circle ratio, and yet it appears throughout mathematics, physics, statistics, engineering, architecture, biology, astronomy, and even the arts. Pi lies hidden in the rhythms of both sound waves and ocean waves, ubiquitous in nature as well as in geometry.

There's little doubt that if we understood this number better—if we could find a pattern in its digits or a deeper awareness of why it appears in so many seemingly unrelated equations—we'd have a deeper understanding of mathematics and the physics of our universe. But the number has always held its cards tight to its chest, ceding little ground in the battle for human comprehension.

THE JOY OF π

This book is a celebration of pi and the spirit of the men and women who have calculated, memorized, philosophized about, and expounded upon this number throughout history. It's an exploration of both the lore and the lure of a number whose place has been cemented in the annals of mathematical fascination by the greatest mathematicians of all time, as well as by hundreds of mathematical crackpots such as those who have fixated on solving the ancient problem of "squaring the circle."

You will find here both a detailed history of the calculation of pi and an overview of the many ways in which pi has crept into our popular culture, including all sorts of ways people have tried to memorize the digits of pi. You'll even find 1 million decimal digits of this famous number for you to peruse, and, if you dare, attempt to make sense of.

The story of humankind's search for pi is filled with history and drama, comedy and misfortune, and—perhaps most of all—a particular, almost peculiar, attention to detail. Perhaps architect Mies van der Rohe was right when he said that God is in the details, for mathematicians have plumbed deeper and deeper within pi's digits with a religious fervor, hoping to find even a hint of understanding. "God ever geometrizes," wrote Plato. And yet, now that we've calculated billions of digits of pi on the world's fastest supercomputers and still found nothing, we must begin to ask: How detailed must we get before we start to appreciate the mystery of the number?

A History of Pi

Early History: 2000 – 500 B.C.E.

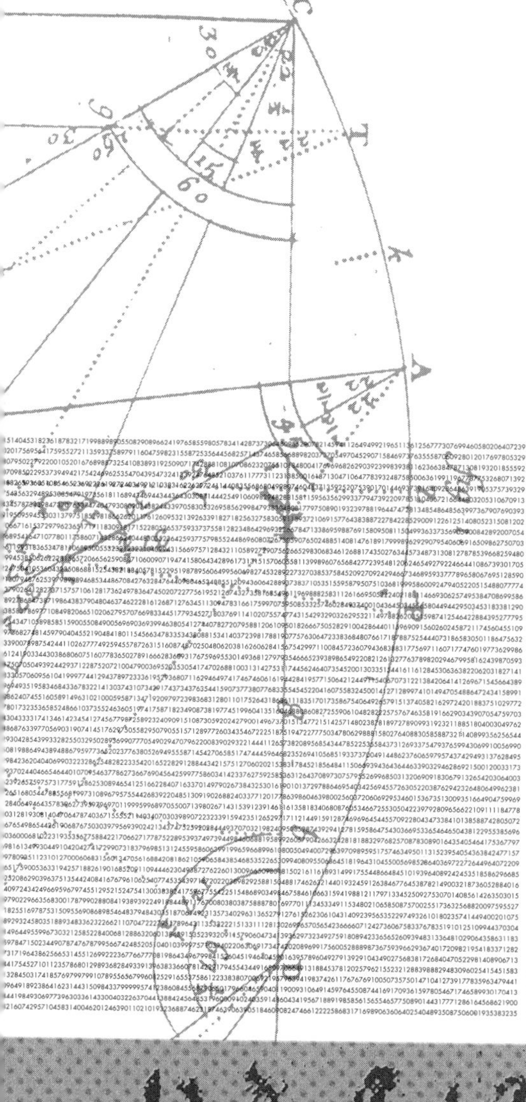

The computation of π is virtually the only topic from the most ancient stratum of mathematics that is still of serious interest to modern mathematical research.

—**Len Berggren, Jonathan Borwein, and Peter Borwein,** *Pi: A Source Book*

It's impossible to know when the first person observed that as a circle got larger, its diameter and circumference grew in constant proportion to each other. Moreover, the area of the circle was also based on this proportion. But what was that constant value?

Simply by experimenting, mathematicians in early civilizations must have figured out that a rope wound around the periphery of a circle equaled just over three lengths across its diameter. With more accurate measurements, they probably discovered the value of the additional bit of rope at more than one-eighth of a length and less than one-fourth.

The earliest-known record of this ratio was written by an Egyptian scribe named Ahmes around 1650 B.C.E. on what is now known as the Rhind Papyrus. He wrote, "Cut off 1/9 of a diameter and construct a square upon the remainder; this has the same area as the circle." Given that we know that the area of the circle equals πr^2, if that area is the square of 8/9, then the papyrus implies that the ratio of circumference to diameter equals 256/81, or 3.16049 . . . .

Ahmes's value was off by less than 1 percent from the true value of approximately 3.141592, showing considerable accuracy for his time. Nonetheless, history shows that his word did not spread. A thousand years later, the Babylonians and early Hebrews were simply using the value 3 for the ratio between circumference and diameter, which wasn't even close.

The formulas on the Rhind Papyrus are also the first recorded instance of someone attempting to "square the circle"—that is, to build a square with the same area as a circle—making circle squaring one of the most ancient mathematical problems in the world and one that has resurfaced time and time again throughout history.

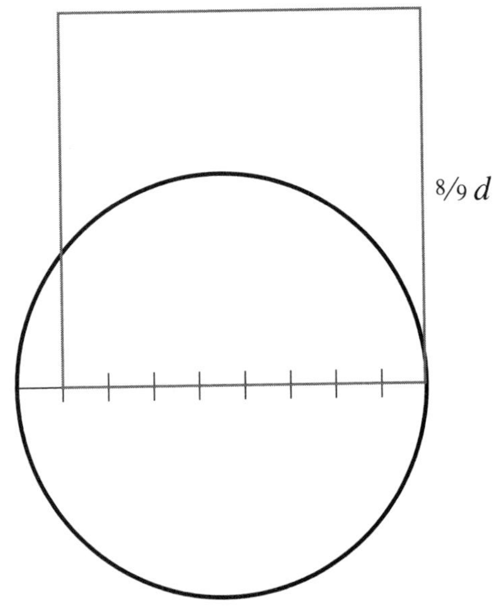

847779777771841566927112240044551677054193107303883694387978890462421759004666609104016362270
064506716725632981356191695775856133339039827759965225099040387093278986221595494379970630648
070964941770800581227055933091621184400463589378563243586419106154006820478790162140445787717
398102952607173000991217971137542433348822666186718059345350035979406261016945589487985287382
394619259273830058657862369012719296382659263937819596877763449192781383915273468510317128350
116775412896963401763368803347613242500654794483551600242312566460801078670258603760993908004
5175626000906555541309840427357430050066877433135282061707299033893705322254670042058896404652
393614283079185404169666783240709559587709423200941561095553433443491385438840861082486242898
596197412565717042400678767123685935372271091567040606219434726020409941395472013175524491583
459442749191929135023558344041872069743458860538337018589765720622546686389914740617138440911
140542444892418125280587378444375990337027144323207852046419314755947583142919416971906297697
904498821308019258759048587570102804989009276466743181741931273879879190866705646017414104518
365473639211201831272642135299075075316741860411390850791740441726580092889664003508561829937
247213684341314956929570401981300160608754127957466419031797332593924021074167670242353517422
118285715161832976814222607309029636948713308807785556663239728334225270656507307251890309039
502297514554508141344442816541436441049217506227064362861017571711204836658149705824635780075
504562645374462805259328415678857985069010580452797562628572208304783543668131330317233238135
264707525779523301528916639528654318999573174578016782672814602226403818995669379948424210982
489742008823311140013410440951609308313090546550315955515473977480221462406761105271613757998
628354396966578355245670079360497518767958500434785944448348747345525999632392588201044528789
576723339110852081379948423471531895261281875108905121354654969246066557673451857177405113980
900507493228007094056920655442879928976809138532882392312742649637907197997852490903046095850

20328130118819189
79875386127705098
31126796867211781
00609428860334160
74080204485324414
14457945472105469
89291664998194159
97508117083997585
52531253479307072
37719482373833676
06554185021133373
57535711604984088
63069626498901511
55629827792230433
34984493678391519
85626850432520084
47985546296912629
97883130293630646
33704503315526375
20404739224157072
85777998802963532
89006984007678189
69756354021766194
29442475372564984
57226550678735909
34057238979378191
46080319827113924
80497941100492298
14317594991993103
28089795747253376
81460617454331326
44892480370134626
42669263173424435
74270517747565067
55634133360059178
31376373759203890
20426517169186542
24484136094659863
62677543332217290
97280677212181229
45017766423271673
31091927523377242
45059080858927565
56434411548443888
95321327028560540
60064352403401177
43942638314926942
03667677312492334
46046152792228711
74737320682073795
04077538070248751
39736974872208079
41836272457926715
86058756843738256
96601528845015936
36057997648795546
66897963172764144
58267140983930260
58444373118321952
73578242992380530
46097925321759529
43764669671785995
65547975260104811
16090019255987703
16036928905354642
19693617900985472
07855125725975325
76878034184282394
19301167647127802
00132449054170687
19406087486425099
24900412437779902
56723623875213448
75468680180570026
77165904571741675
09358537466327846
61471702229127753
81648935814037542
04131631796620462
60168300583584027
80842508470610285
63214676349216441
55965653995124411
15272252098855763
18078808628374443
95373363866289139
43990459186935629
79931691320743108
49123878269670817
37983802760073283
52371305703835388
12019217807455705
39213125048219766
93406394280234533

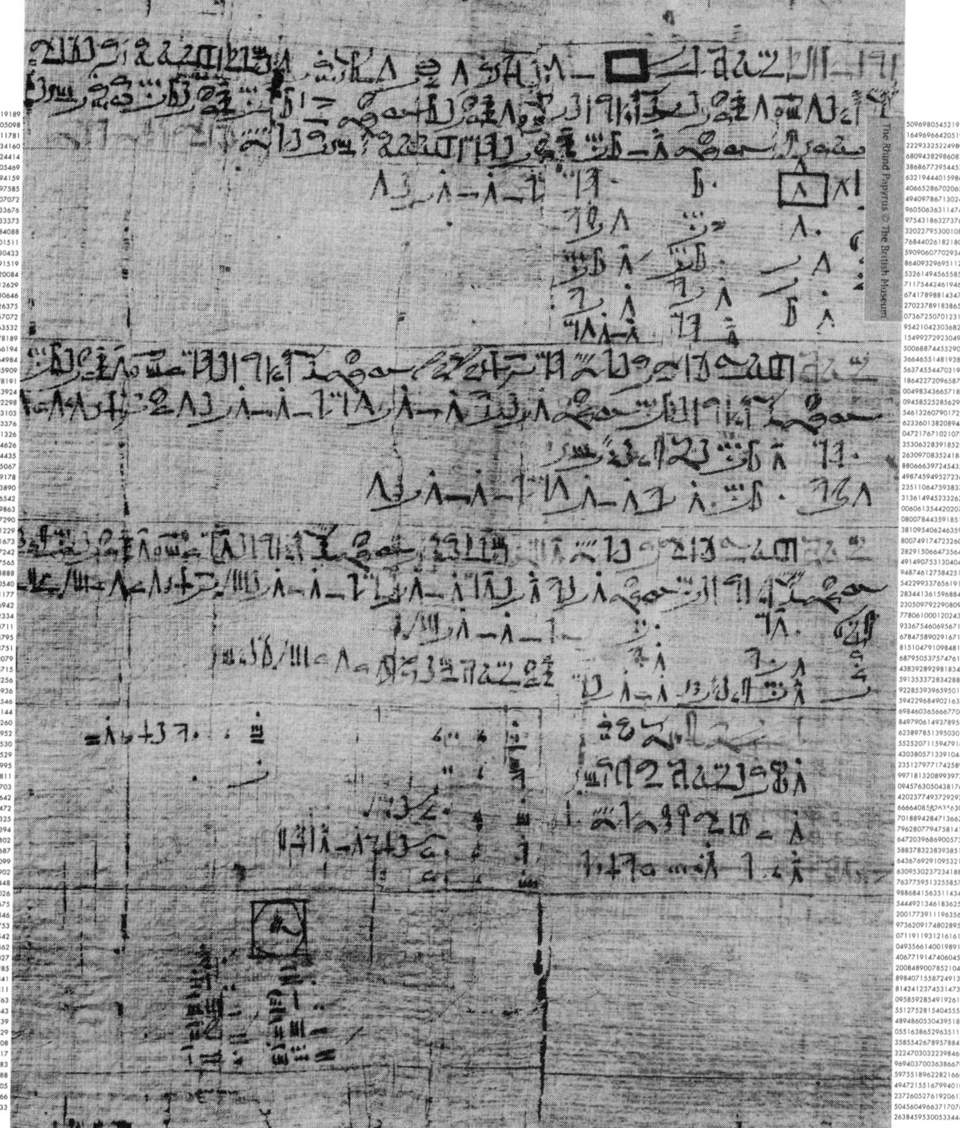

The Rhind Papyrus © The British Museum

50969805452191964
16496966420519223
22293325224980990
68094382986082393
38686773954452673
63219444015986590
40665286702065100
49409786713024508
96050636311474978
97543186327376379
32022795300108771
76844026182180038
59090607702934597
86409329695112338
53261494565585967
71175442461946208
67417898814347780
27023789183865475
07367250701231645
95421042303682530
15499272923049706
50068874455290889
36646551481938480
56374554470319717
18642272096587063
00498343665718303
09458525285629892
54613260790172374
62336013820894764
04721767102107836
35306328391852894
26309708352418426
88066639724543519
49874594952723688
23511064759383705
31361494523326299
00606135442020790
08007844359185142
38109540624635928
80074917472326014
28291506647356469
49149075313040411
94874612758425172
54229933765619132
28344136159688451
23050979229080934
77806100012024307
93367546069567188
67847589029167162
81510479109848199
68795053757476103
43839289298183475
59135337283428884
92285393965950164
59422968490216357
69846036566677068
84979061493789586
62389785139503019
55252071159479162
43038057133910441
23512797717425894
99718132089939724
09457630504381765
42023774937292923
66664085826356304
70188942847136621
79628077947581410
64720396869005733
58837832383938515
64367692910953212
63095302372341887
76377595132558571
98868415635114346
54449213461836258
20017739111963565
97362091748028951
07119119312161615
04935661400198915
40677191474060450
20084890078521044
89840715587249131
81424123745314739
09585928549192619
55127528154045555
48948605304395183
05516386529635114
35855426789578843
32247030322398462
96940370036386670
59755189622821668
49472155167994010
23726052761920617
50456049663717076
26384595300533444

The famous ratio of the circumference to the diameter wasn't named or even given a symbol until three thousand years after Ahmes first wrote about it in the Rhind Papyrus. Mathematicians would always write out phrases such as quantitas, in quam cum multiplicetur diameter, proveniet circumferentia ("the quantity which, when the diameter is multiplied by it, yields the circumference").

The Rhind Papyrus (c.1650 B.C.E.) was first translated and explained in Eisenlohr's Ein mathematicsches Handbuch der alten Ägypter (1877). It is currently in the permanent collection of The British Museum.

Math historians often state that the Egyptians thought $\pi = (256/81)$. In fact, there is no direct evidence that the Egyptians conceived of a constant number π, much less tried to calculate it. Rather, they were simply interested in finding the relationship between the circle and the square, probably to accomplish the task of precisely measuring land and buildings.

The Great Pyramid at Giza has a fascinating relationship inherent in its structure: The ratio of the length of one side to the height is approximately $\pi/2$. Egyptologists and followers of mysticism have supposed for centuries what this means and why it is, for the approximation is significantly better than the value for pi that the ancient Egyptians are understood to have known. However, Herodotus wrote that the pyramid was built so that the area of each lateral face would equal the area of a square that had one side as long as the pyramid was tall. It can be shown that any pyramid with this framework will automatically approximate pi.

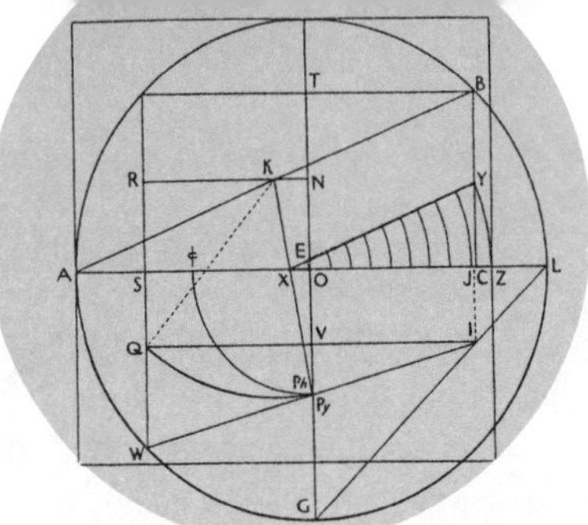

Pi and the Bible

The Bible is very clear on where it stands regarding pi. In the Old Testament, 1 Kings 7:23, we read about the altar built inside the temple of Solomon:

> ALSO HE MADE A MOLTEN SEA OF TEN CUBITS FROM BRIM TO BRIM, ROUND IN COMPASS, AND FIVE CUBITS THE HEIGHT THEREOF; AND A LINE OF THIRTY CUBITS DID COMPASS IT ROUND ABOUT.

This passage (which is nearly identical to 2 Chronicles 4:2), indicating that the ratio of the circumference to the diameter equals thirty cubits divided by ten cubits, or 3, was probably written around the sixth century B.C.E. (though it describes the temple built in the tenth century B.C.E.), and it has troubled mathematicians and scholars for years because it is so far from the truth. Every imaginable explanation for the discrepancy has been proposed, from "This is proof that the Bible is false," to "This is proof that pi really does equal 3, and scientists are lying to us."

Here are several additional dubious rationalizations.

- Solomon's vessel was actually shaped "like a brim of a cup" (2 Chronicles 4:5). The diameter stated in these passages was measured at the top of the cup, while the circumference was measured around the bottom.

Rabbi Moshe ben Maimon (1135–1204), more commonly known as Rambam or Maimonides, wrote, "The ratio of the diameter of a circle to its circumference cannot be known . . . but it is possible to approximate it . . . and the approximation used by scientists is the ratio of one to three and one seventh. . . . Since it is impossible to arrive at a perfectly accurate ratio . . . they assumed a round number and said, 'Any [circle] which has a circumference of three fists has a diameter of one fist.' And they relied on this for all the measurements they needed."

The value for pi in the text is a perfectly adequate approximation for ritual practices the layperson would need to perform. The true value must be culled not only from the words but also from the numerological equivalents of the Hebrew letters: The word circumference is written using the letters Qof, Vaf, He but is read as Qof, Vav. If you look at the numerological equivalents for these two spellings—where each letter in a word equals a number and a word's "value" equals the sum of its letters—you find 111 and 106. The result of dividing these two numbers, then multiplying by the "lay" value of 3 is, surprisingly enough, 3.14150943 . . . .

The brim around this enormous vessel was a "handbreadth" (2 Chronicles 4:5). The diameter included the width of the brim, while the circumference was measured within the brim (implying that the "handbreadth" equaled about one-fourth of a cubit).

Circles are to one another as the squares on their diameters.

—**Euclid**, *Elements*, 12.2

After Ahmes the Scribe recorded his formulations in Egypt, over a thousand years passed before anyone gave much thought to the relationship between circles and squares. Throughout this time, the Egyptians and Babylonians found their basic understanding of the ratio to be adequate for measuring land and building structures. However, the Greeks of the fourth century B.C.E. reexamined the topic of circular measure more closely. They were interested—some say obsessed—not with measuring land but with exploring ideas. For the first time in history, people stopped asking "how much?" and started asking "how come?" It was a golden era of thinking, and while the ratio between circumferences and diameters was not, by far, the most important issue of the day, it was most certainly a focus of some of the greatest minds in ancient history.

The first Greek to attempt to find a definitive relationship between a circle and a square was Anaxagoras of Clazomenae (500-428 B.C.E.). Plutarch wrote that while in prison for teaching that the sun was not a deity (certainly not the first nor last clash between science and religion), Anaxagoras developed a method for drawing a square of equal area to a circle—though, conveniently, Plutarch didn't go on to explain how he did it.

Soon after, Antiphon and Bryson of Heraclea, both contemporaries of Socrates (469–399 B.C.E.), attempted to find the area of a circle using a brilliant new idea: the principle of exhaustion. If you take a hexagon and double its sides and then double them again, and keep doubling them, sooner or later (they figured), you'll have so many sides that the polygon will be a circle.

First, Antiphon estimated the value of the area of a circle by inscribing a polygon in a circle and then calculating the area as each successive polygon came closer to being a circle. Then Bryson took a second revolutionary leap: He calculated the areas of two polygons—one inscribed in a circle and one circumscribed around a circle. He figured that the area of the circle had to fall between the areas of the two polygons—probably the first time that a result was determined using upper and lower boundaries.

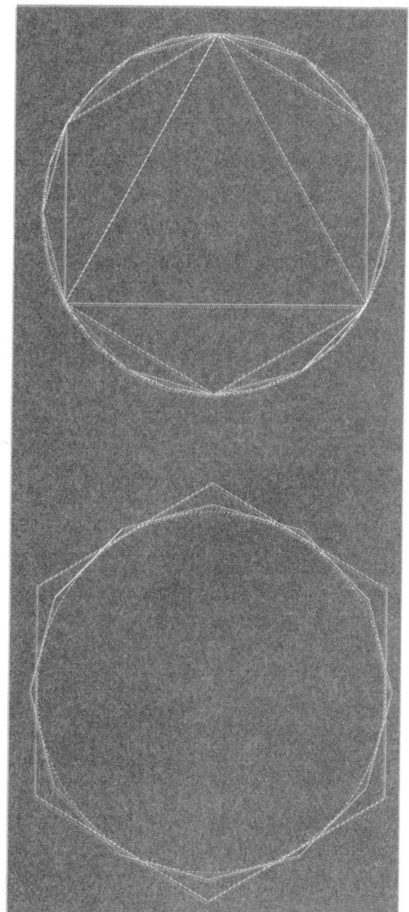

These calculations—which involved finding the areas of hundreds of increasingly tiny triangles—were extremely complex for the time, and neither mathematician was able to calculate many digits.

Some two hundred years after Antiphon and Bryson, Archimedes of Syracuse (287–212 B.C.E.) took up the challenge. Archimedes is perhaps most famous for discovering the principle of buoyancy while taking a bath, prompting him to jump up and run naked through the streets, shouting "Eureka!" But Archimedes was also an extraordinary inventor, mathematician, and physicist. (He discovered the secrets of the pulley and the lever, among other things.) He was without a doubt one of the greatest thinkers in history.

When he turned his attention to circles, he used the exhaustion methods of Antiphon and Bryson in his calculations. However, Archimedes focused on the perimeters of the two polygons rather than their areas, and so found an approximation of the circle's circumference.

He doubled the sides of two hexagons four times, resulting in two 96-sided polygons, and then calculated their perimeters. He later published his findings in his book The Measurement of the Circle: "The ratio of the circumference of any circle to its diameter is less than $3\,{}^{1}/_{7}$ but greater than $3\,{}^{10}/_{71}$." Archimedes knew that he could only describe the upper and lower limits to the ratio, but if you average his two values, you get 3.1419—less than three ten-thousandths from the true value.

However, even more astonishing than Archimedes' accuracy was the fact that he calculated these values without the benefit of a symbol for zero, much less any sort of decimal notation. These concepts had not yet been introduced to Western thought, and wouldn't be for hundreds of years to come.

There is some controversy in the history of mathematics as to whether Apollonius of Perga (a colleague of Archimedes, though thirty years younger) or Archimedes himself actually built upon The Measurement of the Circle, calculating the lower limit of the ratio out to 211875/67441 (about 3.1416). No matter who performed the calculations, this is the last recorded value of pi until the famous astronomer Claudius Ptolemy (87–165) used $3\ ^{17}/_{120}$ (about 3.1417) in Alexandria over two hundred years later.

Actually, to be precise (as mathematicians are wont to be), Ptolemy, in his Megale syntaxis tes astronomias, stated that the ratio equals 3° 8′ 30″, in sexagesimal notation. In other words, pi equals 3 + 8/60 + 30/3,600. This value is within .003 percent of the correct value; if you calculated the circumference of a circle with a diameter of 1,000 feet, Ptolemy's value would come within an inch of the correct measurement.

5117393530415841133265471290567398884435930822283033032221165929854191965597971848854238871580891403693016171725700155706148369068127422955027934635226450068693430774820746663687476147620022750181551796978266737414595043
870588723873389632912139573039946630543402891327746816875466950216141246550370091265917983029038873484176139723433945569356600838016094355613785537468920714544233776467196313846465263157010171323583974874665442363027792 8
5419045015666457881859979478971251481140502377690261728979301308065657163121212079142907054215088898379545365916435512334174598794809276941751149031174605522455785458135586702153090077031955655899599746805741613338361641
6911400992334155643868362258664428079403362670105226669361924674723713640905428985205188351003692681879974656470525450682683936264069944223117912997333641066781735915971628983274177288723020526098042487577710069881962403
7291271628455835847840340492436487818337243203716187881493183663213242424242420147187986601290829544902098739959542872139066776982756308916794217401688235876539750420302448986418896369096316270120557681969929154992775142 54
3788129467665083250351267168466448444547240410124528064217832732227760436916102880783588718437100518084017958014108352816351636038053463076389194761501869867367060501475565451912556348547440616202739383503562785615295889
468170169994014332311095287212448270472060546025850066704075791114136827906978686587117792043561148929968719888032590349546258685078645156073721715 2167299893940149412510356465833636887760906588376967418291919
203119456469780942498386090416033096376452927942341930023017400543432258546250947435455170968354369756035650199238514737184926705972332775797911738152474353163342411729845894129107504

When the Greek philosophers found that the square root of 2 is not a rational number, they celebrated the discovery by sacrificing 100 oxen. The much more profound discovery that π is a transcendental number deserves a greater sacrifice. Again mathematics triumphed over common sense. —Edward Kasner and James Newman, Mathematics and the Imagination

The surface of any sphere is four times its greatest circle. —Archimedes (in a letter to Dositheus of Pelusium)

At the height of their empire (27 B.C.E. to 476 C.E.), the Romans often stubbornly used 3 1/8 for pi—even though they clearly knew that 3 1/7 was closer—because an eighth (a half of a half of a half) was much easier for their legions to work with. In fact, one Roman treatise on the subject of surveying contains the following instructions for squaring the circle: "Divide the circumference of a circle into four parts and make one part the side of a square; this square will be equal in area to the circle." This implies that $\pi = 4$. With this in mind, it's quite astounding that the Romans were able to build as many great structures as they did.

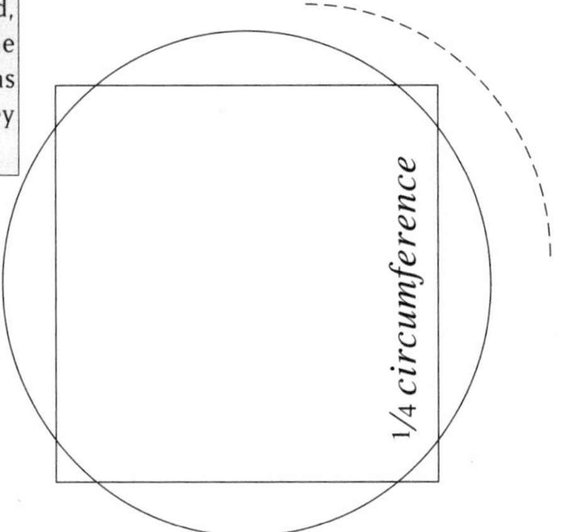

In 212 B.C.E., Syracuse fell to the Roman siege while Archimedes was within the city walls. At seventy-five, he was the greatest mathematician and physicist of antiquity. Plutarch later wrote about his demise: "It chanced that he was alone, examining a diagram closely; and having fixed both his mind and his eyes on the object of his inquiry, he perceived neither the inroad of the Romans nor the taking of the city." When a Roman soldier approached him, Archimedes warned, "Do not touch my circles!" Having no idea who this old man was, and perhaps not even caring, the soldier simply murdered him and went on with his business.

In the Greek alphabet, π is the sixteenth letter (and 16 is the square of 4). In the English alphabet, p is also the sixteenth letter, and i is the ninth letter (the square of 3). Add them up (16 + 9), and you get 25 (the square of 5). Multiply them (16 x 9), and you get 144 (the square of 12). Divide 9 by 16, and you get .5625 (the square of .75). It's no wonder that they say, "Pi are squared!"

Pi in the East: 100 – 700 C.E.

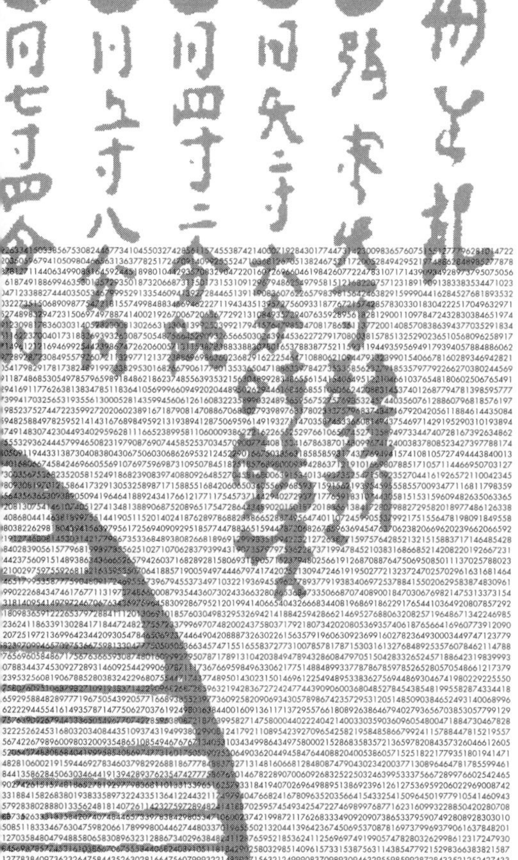

The diameter and the square of half the diameter, when divided by seven and multiplied by twenty-two, are the circumference and the area of the circle.

—Sridhara, *Gaṇitapañcaviṃśi*

While warring governments and religious strife stifled education, research, and the free flow of information throughout Europe during the first millennium, great scientific progress was made in China and India, two of the most important centers of mathematical studies from the second through the eighth centuries C.E.

CHINA

It's long been known that one of the earliest scientific and mathematical civilizations developed in China. However, while Chinese mathematicians were making great strides in mathematics as early as the twelfth century B.C.E., they were still using the value $\pi=3$ in their calculations. China's real achievements in the evolution of measuring circles did not come for another nine hundred years.

Ch'ang Hong was Emperor An-ti's minister and astrologer in the beginning of the second century C.E. Before he died in 139, he wrote that (circumference of a circle)^2 ÷ (perimeter of the circumscribed square)^2=5/8. Using a unit circle (a circle with a diameter of 1), this means $\pi^2/16=5/8$, so that when you work the math, this implies that pi equals $\sqrt{10}$ (about 3.162).

Perhaps because of its visual simplicity, the value $\sqrt{10}$ would become the most popular approximation for pi for years to come throughout Asia, even though it was far from accurate. If you tried to build a circular building 50 feet (15 meters) in diameter with this value, you'd mismeasure the circumference by about 1 foot (30 centimeters) and the area of the floor plan by over 12 square feet (1.1 square meters)!

The Chinese came one step closer to an accurate value for pi when Wang Fau (229–267) stated that if a circle's circumference is 142, then the diameter is 45. It's not clear exactly how he came upon this value, but when divided, the numbers tell us that pi equals 3.156.

Then, over 650 years after the Greeks Bryson and Antiphon approximated the area of a circle using successively larger polygons, the Chinese Liu Hui found the same method and began his calculations. In 263 he published a book in which he demonstrated his method of exhaustion. By inscribing a polygon of 192 sides in a circle, he found that pi lies between 3.141024 and 3.142704. Later, using a polygon with 3,072 sides, he determined that pi equals approximately 3.1416.

However, the real glory must go to the great fifth-century astronomer Tsu Ch'ung-chih and his son, Tsu Keng-chih. Apparently using inscribed polygons with as many as

弧田圖

相較則徵密於約率

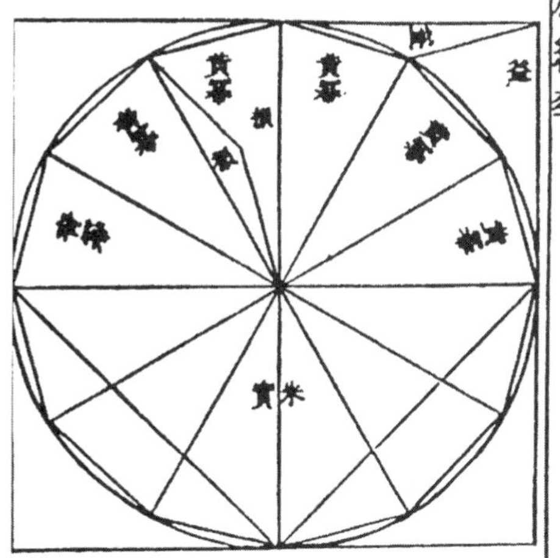

據注意取半圓驗之黃冪令損益相補適滿大方四分之一則青冪適八分之一也合青黃冪爲半外方四分之三朱實與黃冪相等舊以十觚之冪爲圓冪又以半圓論弧矢立法之疎顯然

24,576 sides (they probably started with a hexagon and then doubled the number of sides twelve times), they deduced that pi is approximately 355/113 (about 3.1415929). This is only 8-millionths of 1 percent different from the now-accepted value of 3.14159265 . . . .

The father-and-son team had computed the most advanced calculation of pi in the world, and no one would find a more accurate value for more than a thousand years.

India

India had also developed a mathematical community by the middle of the first millennium, most likely as a result of scientific learning from Greece and Rome first introduced to India with the conquests of Alexander the Great in the fourth century B.C.E. However, news did not travel quickly in those days, and the Chinese value of 355/113 certainly took its time journeying to the West.

In the meantime, Indian mathematicians patiently searched for the mysterious ratio. Around 530 C.E., the great Indian mathematician Aryabhata came up with an equation that he used to calculate the perimeter of a 384-sided polygon, finding it to be $\sqrt{9.8684}$. When it came to publishing his value, however, he approximated the value in the Ganita, a poem in thirty-three couplets. In the fourth couplet, he writes: "Add 4 to 100, multiply by 8 and add 62,000. This is the approximate circumference of a circle whose diameter is 20,000." Sure enough, when you divide 62,832 by 20,000, the result is 3.1416 (ironically, his "approximation" is even more accurate than his square root).

Brahmagupta, the most prominent Indian mathematician of the seventh century, took another tack and mistakenly assumed a pattern could reveal the ratio of circumference to diameter. First, he calculated the perimeters of inscribed polygons with 12, 24, 48, and 96 sides as $\sqrt{9.65}$, $\sqrt{9.81}$, $\sqrt{9.86}$, and $\sqrt{9.87}$, respectively. And then, armed with this information, he made the leap of faith that as the polygons approached the circle, the perimeter, and therefore pi, would approach $\sqrt{10}$.

He was, of course, quite wrong. While it's understandable that he wouldn't have known about Ch'ang Hong's erroneous calculation five hundred years earlier in China, it is curious that he didn't know about Aryabhata's somewhat accurate calculation just a hundred years earlier. Even more curious is that he didn't see that his square roots were converging to a number significantly less than 10 (in fact, the square of pi is just over 9.8696).

Nonetheless, this was the value he expounded, and maybe because $\sqrt{10}$ is so easy to convey and remember, this was the value that later spread from India to Europe and was used by mathematicians around the world throughout the Middle Ages.

Aryabhata wrote that if a equals the side of a regular polygon of n sides inscribed in a circle of unit diameter, and b is the side of a regular inscribed polygon of 2n sides, then $b=\sqrt{[1/2-1/2\sqrt{(1-a^2)}]}$. This is the equation he used to find his well-known sixth-century value for pi.

24605478859944208240180919736271175154047465
63411804862886439875110526018600763208664032
08005880981246682872769158288851453555992972
14513431881771664556450266633627515714226121
27028290235870314678624273023359989513383310
69080367912289757422665249899098159012376515
704327731921850284448811193320110357105719444
38712183523225548677264408667340454413536740
39901046417928811413277329570523323399878009
16026700289290467003455063211355182259645456
36558027046215314706032147678038734544203988
77573153641972943746586782763362311198646746
08317162495938051692232090053533983610528084
87974347050510512429799469695877329008120707
97287965358392324265767339214438047036170652
95956729932344168693092018662571582035045922
27460113349178476867831063630236724355370932
56269498230726186313109105016432061267424608
67916703779309406696071354477720412401713871
52541478713374566022914274536828100929205588
90079508483723267871865955621283765493043122
74644597738111563966740927499199030967831570
44379273964166675109789264093117468241878846
53928794391428071913722819450621119960494201
41675675141552265693285969399005410111647767
52925649440428795835710036845090703458019087
49999309273423323790664741074628981171010402
77883382145098302778833821450983160613718505
84279038953949613459869455343321733883804422
92218684824710117148515834710609975786976196
81601243733023068446927105578932616600129599
34985974917184503344610562408400109524903112
91513102073536606699142509744167108918044279
26385025576622062566434705688881209134312965
478161984539675154821810244160624449

Brahmagupta, in the seventh century C.E., wrote that the "practical" value for pi was 3 and the "neat" value was $\sqrt{10}$.

916398273490435941139589840434895757833246482216524972533181193055585614050
310507654858991552554265288628752889554577367874202977037846847563664249476
708485430735413284036163491347107446832985898095111020124254488464307122717
488658696436722375125740766380757796859385182321580379013885132456704225278
538766101351956828652339460402003567338602520551347530790074689452614361638
124660209433968818299857253465435528540463610413121499376921630261483151823
469420962791154941719466072066552844004435657532664143893427722090557518423
691208034737988670796922839869375088816146073838246420008153936740018862573
073695349973083672528101494304364563497521354531951950035076482370361845384
975636163397442943098863871989880818808674749583176022298467250195918371787
001546471943774402458796441934330527377861745024524970714990700051872692928
345871786309174843878550639754778139797614710297480525893069221666222523537
344901135398626602814192647629370976801318720414066687625545954222924938494
627117755017586202137887676002980515741123780955192781815908206636365403568
683324455662009516046337522568825585458292001930638153387365651794574370258
875626473210773227646623152269937958253816250741193599257543470320751896392
792921623091299025904455121720931896617993469495415021868337701522075911300
888689023857991528263986782465460887462785262268142473318858857241665126195
900032922440472840896196026492377307279303286983507199509173362220690426621
1379357378789633982192711117924375186838175762134729222730484110905289312739
756654644015991089205635953955062268490348177834016388807477585910604735866
456072994009420006312043562308116498145514965551300585461173552405216715566
6041333475875987920445592175677563283767222769011541649211224642236039540336
845501134265424744898954599679203644242966524827350687996495015740462148251
111674016381288237054926766797280057463290619456179973094448723746706306283
461937692637284371025062943023983874718041127794451518210864000155847579128
464012873995097762977082622634588250520781834576050530815712768164616012674
561513103910717697384557873224133003000553471951166901258113520801563037304
690809309792735365856491357471135090441275907649029919388200826217393959286
123336572970664641027058783855131893465796268593304795602601154503596771014
005799333688900402207538482513993086371634336600792371240645761765003641061
220543568868817740625305700602301898291109153407117751712442370364363715890
220116231710263565013024399121540427012730391660434852892171767800544353796
026814476987479405571599377835639966210066927419271468108962040736111634720
258986246474408196120403368752089701088063354284436925218017425121196785699
110583349449991683094494698470780636754666776782538372304052848929117305480
298931061328228524301397442127840108229799225637499186161909539509229235240
387265633496244744690348057513565946504625030962501118599636302403654187824
4570740245894880605074168390715058032424183755862679604489403111842071561842
6638993005968351960880991550054081911160942615617799649455573893623350956021
693845302940741535422017008850593410802153774416896976552390007001131094692
800034443560636076613103027287389231791016021743160036372135135506555681162
767164832287962390037143316348095868924384711690483078965100591104965015992
831438312018932525166768955897310518020709156128212794785768231503099654870
137801420342350862188944511309174155201212503779765726305117588445579181661
243191479349987937188974667677782724332922702482645480284999856755494526946
870327503783940036651442685682081309020949057899622100814077366965566279789
587599381603739294081898326023119790605145978038449412185507347234440464136
333171482978197669866965514005181845419763310556350448849713422360339130058
979717346782373472329230517388505004636025681998062728258112455591586015018
439090409864180971710075461884773934911273571127107533095079036197946170873
344664805241788806067731106455884142874312055368645075413123789205016418245
598529170285529823491756815198174953565040453735880040973693100210161974099
408857233681398906852305802152257830798584444988490026722154928888612925028
852813527173780318207628086658198702133918612113360246187362649128598385704

The Chinese of the first millennium had two mathematical advantages over most of the rest of the world. First, they used decimal notation where most people made do with ratios. Second, they used a symbol for zero. It seems like such an obvious thing to do, yet European mathematicians (and bankers, and engineers) did not use a symbol for zero until the late Middle Ages (the idea was slowly introduced through contact with Indian and Arabic thinkers).

A Millennium of Progress: 600 – 1600 C.E.

This mysterious 3.14159 . . . which comes in at every door and window, and down every chimney.

—**Augustus De Morgan,** *A Budget of Paradoxes*

The first millennium C.E. saw the Dark Ages in Europe, which were filled with war and strife following the breakdown of the Roman Empire and the rise in power of early Christianity. Any budding scientific interest in Europe during these years was effectively quelled by religious intolerance or destroyed by warring factions. But knowledge has a way of traveling to where it will flourish, and pi (along with many other facets of Western thought) managed to pop up in the more nurturing academic climate of the Muslim world.

For instance, by 500 C.E. the astronomical and mathematical discoveries of Ptolemy had filtered from Alexandria into the Sassanid Persian culture, helping to support the astrological and astronomical underpinnings of their state religion, Zoroastrianism. Later, in the seventh century, their empire fell to the Arabs, and Ptolemy's works were incorporated by Arab thinkers who, by then, had already been introduced to many other ideas from India and Greece.

By the ninth century, math and science were thriving in the Arab cultures, especially in current-day Iraq where Abu Abd-Allah ibn Musa Al'Khwarizmi, one of the greatest Arabian mathematicians, lived and taught. It is unclear whether Al'Khwarizmi actually attempted to calculate the ratio of the circumference to the diameter himself (though it would be surprising if he didn't). It is clear, however, that he used the values $3\,{}^{1}/_{7}$, $\sqrt{10}$, and 62,832/20,000 in his works—the first value he attributed to the Greeks, the latter two, to Indian mathematicians. More important, he wrote using the Hindu-Arabic numerals, including the zero and the decimal point. By the end of the millennium, Arab sciences, which incorporated the writings of the ancient Greeks, Hebrews, and Indians, had spread West along the North African coast to the Moors, who by then had conquered most of Spain.

Finally, at the beginning of the second millennium, European texts that had been preserved in Arabic began to filter back to Europe from the Middle East. This was fueled partly by a newly surfacing European interest in mathematics, and partly by widespread superstitious beliefs based on astrology that had generated a hunger for astronomical information. The interest drove soldiers, clergy, and merchants to collect information from the Middle East and bring it to the West. Perhaps most important, it became clear during this time that knowledge of the sciences would bring power to those who held it.

In 1085, the Spanish king Alfonso VI of Castile captured the city of Toledo from the Moors and, with it, an immense library. He arranged for the translation of scientific works from Arabic, Greek, and Hebrew into Latin, reintroducing the works of great thinkers like Aristotle and Plato to western Europe. Similarly, the returning soldiers from the Christian crusades in the eleventh through thirteenth centuries carried back books and teachings along with their copious plundered treasures.

The church was not above intrigue when it came to collecting scientific information. At the beginning of the twelfth century, the English cleric Adelard of Bath disguised himself in order to study as a Muslim in Cordoba, Spain. There he found Euclid's Elements and Ptolemy's Almagest and translated them into Latin so that other scholars in western Europe could read them. He also translated Al'Khwarizmi's works, introducing Arabic numerals and notation to Europe.

After several centuries of increased trade between the Middle East and Western countries, Italy (particularly Venice) monopolized the trade routes in the early part of the second millennium. Leonardo de Pisa (more commonly known by his nickname Fibonacci) was the son of an Italian diplomat stationed in North Africa in the late twelfth century. He learned the Arabic sciences early in life, and later, when he lived as a merchant in Italy, he was clearly as interested in the exchange of information as he was in that of tangible goods. In 1202, at the age of thirty-two, he wrote Liber abaci, which not only contributed to the establishment of Arabic numerals (including 0, thanks in part to the writings of Al'Khwarizmi) in Europe, but also introduced a puzzle that led to the number sequence that he's so famous for today.

In 1220, Fibonacci approximated pi to be 1440/(458 $^1/_3$) or 864/275 (about 3.1418) in his Practica geometriae. Clearly, Tsu Ch'ung-chih's advanced calculations in China eight hundred years earlier had still not traveled far enough to reach Europe, as this value is only about .0001 more accurate than Archimedes'.

While the study of mathematics—and in particular, pi—was far from stagnant throughout the Middle Ages, little actual progress was made, and the values for pi remained less accurate than the earlier Greek, Chinese, and Indian calculations. Albert of Saxony (1316-1390), bishop of Halberstadt, in his De quadratura circuli, wrote that the ratio of circumference to diameter was exactly $3\,{}^{1}/_{7}$. In the mid-fifteenth century, Cardinal Nicholas of Cusa claimed to have squared the circle exactly, finding the ratio of the circumference to the diameter to be 3.1423. Unfortunately, his method was later proved to be false by Regiomontanus (a.k.a. Johannes Müller, 1436–1476).

In fact, it was not until the late sixteenth century that another significant step was taken—this time by a French lawyer and amateur mathematician named François Viète.

In 1579, Viète used the tried-and-true Archimedean method of inscribed and circumscribed polygons to determine that pi was greater than 3.1415926535 and less than 3.1415926537. To achieve this feat, he doubled the sides of two hexagons sixteen times, finding the perimeters of polygons with 393,216 sides each. However, while his value, accurate to 10 places, was the most precise measurement of pi in history, Viète's truly great accomplishment was to describe pi using an infinite product.

in′finite prod′uct, *Math.* a sequence of numbers in which an infinite number of terms are multiplied together.

In his 1593 book Variorum de rebus mathematicis responsorum liber VIII (Various mathematical problems, volume 8), Viète broke down his polygons into triangles and found that the ratio of perimeters between a regular polygon and a second polygon with twice the number of sides equaled the cosine of θ.

sine (sīn), *n.* **1.** *Trig.* **a.** (in a right triangle) the ratio of the side opposite a given acute angle to the hypotenuse. **b.** (of an angle) a trigonometric function equal to the ratio of the ordinate of the end point of the arc to the radius vector of this end point, the origin being at the center of the circle on which the arc lies and the initial point of the arc being on the x-axis. *Abbr.:* sin **2.** *Geom.* (originally) a perpendicular line drawn from one extremity of an arc of a circle to the diameter that passes through its other extremity. **3.** *Math.* (of a real or complex number x) the function $\sin x$ defined by the infinite series $x - (x^3/3!) + (x^5/5!) - + \ldots$, where ! denotes factorial. Cf. **cosine** (def. 2), **factorial** (def. 1). [1585–95; < NL, L *sinus* a curve, fold, pocket, trans. of Ar *jayb* lit., pocket, by folk etym. < Skt *jiyā, jyā* chord of an arc, lit., bowstring]

sine (def. 1)
ACB being the angle, the ratio of AB to BC is the sine; or, BC being taken as unity, the sine is AB

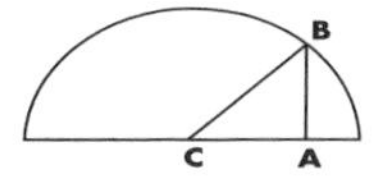

co•sine (kō′sīn), *n.* **1.** *Trigonom.* **a.** (in a right triangle) the ratio of the side adjacent to a given angle to the hypotenuse. **b.** the sine of the complement of a given angle or arc. *Abbr.:* cos **2.** *Math.* (of a real or complex number x) the function $\cos x$ defined by the infinite series $1 - (x^2/2!) + (x^4/4!) - + \ldots$, where ! denotes factorial. *Abbr.:* cos Cf. **sine** (def. 3), **factorial** (def. 1). [1625–35; < NL *cosinus.* See co-, SINE[1]]

cosine
ACB being the angle, the ratio of AC to BC is the cosine; or, BC being taken as unity, the cosine is AC

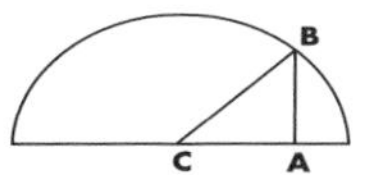

82536534878288646972253655754525228031688471039269429944017744150565410839248185380972246265514689030
00207821275494505279154369817549666187513478319186574125583553974077373341601561144528150171617511799
96399406119110086304704795713409531582791149697550614252596618790194047452875733438892990800297698778
93086987339902732472293618765193297280946392158801058120917330356069608825523217970057600041590448817
93922984453580379746071294707608200651516833656451241281129400207911460827424310952033600285057851031
78129690111967320860994900367426065788332367599890317467841882737621128226150437357826128239235832602
35062150253881205038087610757792341102006638835964616931815750428606621212402530812757970025787253844
90578740240767751761182828022057007680331731437332224972089532223176991483092291852524674905187716587
1 9 2 8 3 3 9 1 4 5 1 2 7 2 2 2 4 3 9 1 0 7 6 8 8 0 3 8 1 4 6 7 6 4 7 7 6 8 2 5 6 0 5 1 9 9
16242899466641586857003358971005817274611700131317272015264539575067017238873314438527194969975372458
50451851121245323760092430472399543989332763258474199660212612529805661849768230504120572683027889590
12984790370101236347773267203908107116393032892689698589802760428530981257919573240805314535999506802
81647637678616204990872205057179263264470138021032744757850959815376927943735399559906920110868457276
15873747414978132199221009794636168368837698006801932672463563313936198022844660290825749708876116126
19391798898914147203605599369088389305805359369339114503166658376790682533810154946336850527021605286
58989694225709635345492408795324498345015230231036833493083408235168291518964166715750476290195346765
50504543318915726570514987763841490791267283803179053794039065513432425793133041324948076088104697312
49545345457856264329245753975443631106604365289403443842934131029921856386196903953622936190101639935
28535010572993277183944687864902771924119694776679674321691661740183719065604639000765211961148350720
75559291017853787705695420746007253475463298759180083020271502977497891528398945332554071951666575323
09264951394211425540451153778645696623468005001055766568622256559753200069485364386223037984856936822
38749031954900491665783397436698609183391998723719472588452887254012846450566305472362710992642785702
45829223730422001039892514376074181119767998004961158488903136574404814727769349793351969079124128680
4
95050177445358305674042673285789757264025168112914401728938893906007862203398066619650785808534824907

With this identity in hand, he used the half-angle formula and found a way to describe pi as a product:

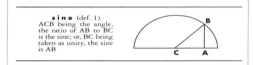

$$\frac{2}{\pi} = \sqrt{\frac{1}{2}} \times \sqrt{\frac{1}{2} + \frac{1}{2}\sqrt{\frac{1}{2}}} \times \sqrt{\frac{1}{2} + \frac{1}{2}\sqrt{\frac{1}{2} + \frac{1}{2}\sqrt{\frac{1}{2}}}} \times \cdots$$

This was, in fact, possibly the first time anyone had used an infinite product to describe anything, and it was one of the first steps in the subsequent evolution of mathematics toward advanced trigonometric identities and calculus. Unfortunately, while the equation was a breakthrough, Viète found that it was of little use when it came to actually calculating pi—it takes too many iterations of calculating complicated square roots to achieve even a small number of digits.

Three other late-sixteenth-century mathematicians employed the Archimedean method of polygons to calculate pi. First, in 1585, Dutch mathematician Adriaen Anthonisz used inscribed and circumscribed polygons to show that $377/120 > \pi > 333/106$. (In decimal notation, this means $3.14167 > \pi > 3.14151$.) By itself, this is not that impressive; however, as luck would have it, when he averaged the numerators and denominators, he independently rediscovered the Chinese value 355/113. Next, eight years later, another Dutch mathematician, Adriaen Romanus found pi to 15 places, using an inscribed polygon that had over 100 million sides!

Finally, Ludolf van Ceulen spent years calculating pi to 20 places using the same method that Archimedes used, except that his polygons had more than 32 billion sides each (60×2^{29}). By the time he died in 1610, van Ceulen had calculated 35 digits of pi using the same methods mathematicians had used for thousands of years. His feat was surely one of stamina and patience, if nothing else, and when he died, legend has it that his value for pi was engraved on his tombstone in St. Peter's churchyard in Leyden, Germany. (The stone has unfortunately been lost.)

However, just over ten years after van Ceulen died, there was a breakthrough in the search for pi that revolutionized the way that the mysterious ratio of circumference to diameter was calculated and that propelled mathematics toward greater heights. Ironically, this breakthrough quickly made van Ceulen's years of work entirely obsolete.

In Germany today, pi is still sometimes referred to as the Ludolfian number (*die Ludolphsche Zahl*), in honor of Ludolf van Ceulen, the sixteenth-century mathematician who spent much of his life calculating 35 digits of pi.

The eleventh-century Michael Constantine Psellus, known by his Byzantine contemporaries as Philosophon hypatos, a Prince of Philosophers, was clearly a forward thinker of the time. Unfortunately, mathematics was not his forte. Apparently, his favorite way to find the area of a circle was to take the geometric mean between an inscribed and circumscribed square. The result fixes pi at the square root of 8, or 2.8284271 . . . .

In 1475, Piero della Francesca, an Italian painter, published *De corporibus regularibus*, in which he noted that $\pi = 22/7$.

Leonardo da Vinci (1452–1519) and the artist Albrecht Dürer (1471–1528) each briefly worked on the problem of squaring the circle, but neither offered any new insights. In fact, in 1525, Dürer simply used the value 3 1/8 in his *Underweysung der messung mit dem zirckel und richtsheyt* (Instructions for mensuration with compass and straightedge). However, Leonardo da Vinci did invent a novel way to find the area of a circle: If you have a wheel whose thickness is half its radius, and roll it one full revolution, the area of the track it leaves equals the area of the wheel. The track, you see, is a rectangle with a width of $\frac{1}{2}r$ and a length of $2\pi r$; multiply the two, and you get πr^2.

Georg Joachim von Lauchen (1514–1576), often called Rhäticus, published a set of trigonometric tables. In these tables, he states that sin 10″ equals .000048481368092. Because half a circumference equals 648,000″, you can multiply sin 10″ by 64,800 to obtain 3.141592652 (accurate to 8 places). Whether Rhäticus actually performed this calculation is unknown, but the tools were there to do it.

Breakthroughs in Mathematics: 1600 – 1900 C.E.

Ten decimals are sufficient to give the circumference of the earth to the fraction of an inch, and thirty decimals would give the circumference of the whole visible universe to a quantity imperceptible with the most powerful microscope.

—**Simon Newcomb, quoted in *Mathematics and the Imagination* by Edward Kasner and James Newman**

The three hundred years from the tail end of the Renaissance through the Victorian Age were an extraordinary time in the history of mathematics. It was as though a seed, after two thousand years of dormancy, had finally found in western Europe the proper climate to germinate, grow, and—by the nineteenth century—bloom into a glorious flower. Some of the most fascinating and insightful mathematicians of the second millennium appeared at this time, each preparing the ground for the next, in a series of steps that revolutionized mathematics and scientific thought. And as mathematics evolved, so did the search for and study of pi.

The brute force approach to calculating pi began losing its appeal by the early 1600s. The exhaustion method's laborious multiplications, divisions, and square roots regarding areas of polygons with billions of sides, which occupied most of Van Ceulen's final years, were too cumbersome to be attempted by many others. Humankind would either have to find a better way to calculate pi or be stuck knowing only a handful of digits forever.

There is a famous formula—perhaps the most compact and famous of all formulas—developed by Euler from a discovery of De Moivre: $e^{i\pi}+1=0$. Elegant, concise and full of meaning. . . . It appeals equally to the mystic, the scientist, the philosopher, the mathematician. —Edward Kasner and James Newman, Mathematics and the Imagination, 1940

In 1621, Dutch mathematician Willebrod Snell found a way to work smarter rather than harder. Where his predecessors had doubled the sides of a polygon in order to gain a more accurate approximation of a circle's perimeter, Snell found a way to calculate a better approximation using the same number of sides.

By simply inscribing and circumscribing a hexagon on a circle, he was able to determine that pi was between 3.14022 and 3.14160. This was closer than Archimedes' approximation using a 96-sided polygon! By doubling the polygon's sides only four times and using a 96-sided polygon, Snell was able to find pi accurately to 6 digits, and

Archimedes approximated the perimeter of a circle by measuring the sides of inscribed and circumscribed polygons. However, it turns out that there are better ways to approximate the length of an arc in a circle than using polygons. Willebrod Snell still measured arcs using an upper and lower boundary, but was able to find much more precise boundaries by using other geometric methods. In the figure, you can see how Snell approximated the length of arc PY to be less than PB_1 and greater than PB_2. (Point Y is stationary and point A moves, depending on which segment is the length of the radius.)

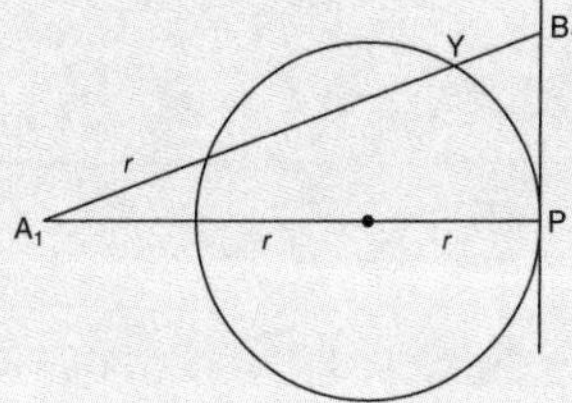

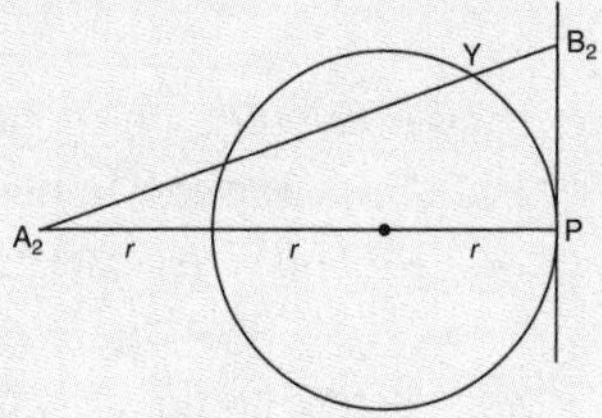

with a little more work was able to verify Van Ceulen's 35 digits. Nonetheless, while he believed firmly in his theories, he was never able to prove them.

Three years after Snell died, in 1626, a man named Christiaan Huygens was born. Like Snell, Huygens was Dutch, and, like Viète, he was trained as a lawyer. In fact, he didn't take up math until he was in his twenties, but he attacked the field with a vengeance and soon after found a rigorous proof for Snell's theorems. What's more, he improved on the theories significantly, so that by simply inscribing a triangle he was able, incredibly, to match Archimedes' approximation for pi; with a hexagon, he was able to find 9 accurate digits, using the limits 3.1415926533 and 3.1415926538.

Neither Snell nor Huygens was interested in calculating a record number of digits; they just wanted to be able to calculate it more efficiently. And although they laid the groundwork for a new analytical approach to "squaring the circle," they were, in fact, the last two great mathematicians who focused on an Archimedean solution using polygons. This is because around the same time Huygens was working with polygons, the stage was being set for integral calculus.

An English contemporary of Huygens, John Wallis was a mathematician and cryptographer with a fresh approach to finding the area of a circle: He looked at approximating the area of a quarter circle using infinitely small rectangles. He did not have the benefit of the notation or identities that calculus would have afforded, as it had not yet been invented, but in 1655 he was able to painfully eke out the formula that to this day bears his name.

John Wallis's formula

$$\frac{\pi}{2} = \frac{2 \times 2 \times 4 \times 4 \times 6 \times 6 \times 8 \ldots}{1 \times 3 \times 3 \times 5 \times 5 \times 7 \times 7 \ldots}$$

Like Viète's, Wallis's equation is an infinite product, but it is different in that it only involves rational operations with no need for messy square roots. Wallis's formula is also interesting because it slowly converges on pi—the first term is higher than pi, the second term is lower, the third is higher again.

There were many other great mathematicians living in the seventeenth century—Blaise Pascal, Johann Kepler, Bonaventura Cavalieri, and Pierre Fermat, to name a few. Each contributed an important piece of the puzzle and got one step closer to the breakthrough of calculus.

One such mathematician was the Scotsman James Gregory. Before his untimely death at age thirty-six in 1675, he found an extremely elegant solution to calculate arctangents, which then led to a completely new way to calculate pi: arctangent series.

Gregory's series

$$\arctan x = x - \frac{x^3}{3} + \frac{x^5}{5} - \frac{x^7}{7} + \frac{x^9}{9} - \frac{x^{11}}{11} \ldots$$

Three years after Gregory came up with this new solution, German-born Gottfried Wilhelm Leibniz independently found the same arctangent series and published it along with an important "special case": by inserting the number 1 into the series, you can easily approximate $\pi/4$.

Nonetheless, while the Gregory-Leibniz series is impressive in its elegance and simplicity and for what it tells us about the nature of pi, it is lousy when it comes to actually calculating digits. It would take 300 terms of the series to obtain just 2 decimal digits of pi, and thousands more to become actually useful.

If the name Leibniz means anything to you, it's probably because he was one of the fathers of calculus. The other father was, of course, Isaac Newton (1642–1727). Newton's brilliance is renowned, but his involvement with pi is not. In 1665, while the plague was ravaging London, Newton retreated to Woolsthorpe and spent his days contemplating calculus. He achieved many great works during this time, including finding at least two infinite series

The Gregory series was found in 1671 and heralded a new age in calculating pi. The reason: The tangent of 45 degrees is 1, 45 degrees equals $\pi/4$ in radians, and therefore if you take the arctangent of 1, you get $\pi/4$ radians. Simply by substituting the number 1 in the series, you can calculate one-quarter of pi.

$\pi/4 = 1 - 1/3 + 1/5 - 1/7 + 1/9 - 1/11 \ldots$

While it would be astonishing if Gregory had not tried substituting 1 to find this simple formula for pi, none of his correspondence includes such a calculation.

arc′ tan′gent, *Trig.* the angle, measured in radians, that has a tangent equal to a given number. *Abbr.:* arc tan; *Symbol:* $\tan^{-1}$ Also called **inverse tangent.** [1905–10]

In order to obtain 100 digits of pi using the Gregory-Leibniz series, you would have to calculate more terms than there are particles in the universe.

for pi. Calculating pi was such a simple task for a thinker of Newton's stature, and yet the number consumed hours upon hours of his time. He later wrote in correspondence, "I am ashamed to tell you how many figures I carried these computations, having no other business at the time."

Clearly, finding the ratio of a circle's circumference to its diameter was no longer just a question of basic calculations. Calculus and arctangent series enabled mathematicians to make much faster calculations than measuring polygons; in fact, calculating just four terms of one of Newton's series yields 3.1416. The issue soon became efficiency—who could come up with an equation that converged most quickly on pi.

With these new tools in hand, the search for digits of pi took a sudden leap at the end of the seventeenth century. In 1699, Englishman Abraham Sharp found 72 digits with the Gregory-Leibniz arctangent series. In 1706, another Englishman, John Machin, used the difference between two arctangents to find 100 digits of pi. Then, perhaps spurred by a competitive spirit, French mathematician Thomas Fantet de Lagny calculated 127 digits in 1719 by using the same technique as Sharp. However, seventy-five years later, Austrian mathematician Georg Vega calculated 140 digits of pi and found that de Lagny had only found 112 correct digits.

Sir Isaac Newton found several infinite series for pi including this one:

$$\frac{\pi}{6} = \frac{1}{2} + \frac{1}{2}\left(\frac{1}{3\times 2^3}\right) + \frac{1\times 3}{2\times 4}\left(\frac{1}{5\times 2^5}\right) + \frac{1\times 3\times 5}{2\times 4\times 6}\left(\frac{1}{7\times 2^7}\right) + \cdots$$

Abraham Sharp's equations were based on the fact that arctan√(1/3) equals π/6 (that is, the tangent of 30 degrees is √(1/3), and 30 degrees is π/6 radians). Then, given the Gregory-Leibniz series:

$$\frac{\pi}{6} = \sqrt{\frac{1}{3}} \times \left[1 - \left(\frac{1}{3 \times 3}\right) + \left(\frac{1}{3^2 \times 5}\right) - \left(\frac{1}{3^3 \times 7}\right) + \cdots\right]$$

$$\pi = 2\sqrt{3} \times \left[1 - \left(\frac{1}{3 \times 3}\right) + \left(\frac{1}{3^2 \times 5}\right) - \left(\frac{1}{3^3 \times 7}\right) + \cdots\right]$$

Then, in the middle of the eighteenth century, one of the greatest and most prolific mathematicians of all time briefly turned his attention to calculating pi. Leonhard Euler (pronounced "oiler") was born in Switzerland in 1707 and worked throughout Europe during his life. By the time he was thirty, he was blind in one eye; by the time he was sixty-five, he had lost his sight completely. Nonetheless, he continued to work, dictating his copious theories and calculations to assistants.

Euler found many arctangent formulas and infinite series to calculate pi, including a method using arctangents that converged much faster than Gregory's, which he used to calculate pi to 20 places in a single hour.

After Euler's brilliant steps, the nineteenth century looks positively sparse when it comes to breakthroughs in methods for calculating pi. In fact, it wasn't until the beginning of the twentieth century that another mathematician would come up with a new set of equations for the problem.

When John Machin calculated 100 digits of pi, he used π/4 = 4 arctan(1/5) - arctan(1/239). This formula turns out to be particularly useful for calculating pi because arctan(1/5) is easy to calculate using the Gregory-Leibniz series, and arctan(1/239) converges very quickly.

Euler's formulas:

$$\frac{\pi}{4} = 5\arctan(\frac{1}{7}) + 2\arctan(\frac{3}{79})$$

$$\frac{\pi}{4} = 2\arctan(\frac{1}{3}) + \arctan(\frac{1}{7})$$

$$\frac{\pi^2}{6} = \frac{2^2}{(2^2-1)} \times \frac{3^2}{(3^2-1)} \times \frac{5^2}{(5^2-1)} \times \frac{7^2}{(7^2-1)} \times \frac{11^2}{(11^2-1)} \times \cdots$$

$$\arctan x = \sum_{n=0}^{\infty} \frac{2^{2n}(n!)^2}{(2n+1)!} \times \frac{x^{2n+1}}{(1+x^2)^{n+1}}$$

$$\frac{\pi}{2} = \frac{3}{2} \times \frac{5}{6} \times \frac{7}{6} \times \frac{11}{10} \times \frac{13}{14} \times \frac{17}{18} \times \frac{19}{18} \times \frac{23}{22} \times \cdots$$

$$\frac{\pi^2}{6} = \frac{1}{1^2} + \frac{1}{2^2} + \frac{1}{3^2} + \cdots$$

$$\frac{\pi^4}{90} = \frac{1}{1^4} + \frac{1}{2^4} + \frac{1}{3^4} + \cdots$$

$$1 - \sin x = \left(1 - \frac{2x}{\pi}\right)^2 \left(1 + \frac{2x}{3\pi}\right)^2 \left(1 - \frac{2x}{5\pi}\right)^2 \left(1 + \frac{2x}{7\pi}\right)^2 \cdots$$

$$\frac{\pi^3}{32} = \frac{1}{1^3} - \frac{1}{3^3} + \frac{1}{5^3} - \frac{1}{7^3} \cdots$$

Euler also developed an equation that some believe to be among the most fascinating of all time:

$$e^{i\pi}+1=0$$

This equation is only a short step away from his discovery that $e^{ix}=\cos x + i \sin x$ (because $\sin\pi=0$ and $\cos\pi=-1$), and would later become critical in the proof that pi is irrational and transcendental.

Nonetheless, the digit hunters continued to toil away using previous methods to find even more digits. After J. F. Callet published 152 digits in Paris in 1837, William Rutherford used $\pi/4=4\arctan(1/5)-\arctan(1/70)+\arctan(1/99)$ to find 208 digits in 1841. Six years later, Thomas Clausen found 248 decimal digits using both Machin's formula and one of Euler's, and in 1853, Rutherford came back with 440 digits. Unfortunately, William Shanks blew Rutherford's calculations out of the water, finding 607 digits that same year. Finally, by 1873, Shanks had calculated 707 digits of pi. The tragedy was that he had made a mistake after the 527th place, and the following digits were all wrong. But no one knew of his error, and his 707 digits were widely accepted until D. F. Ferguson proved him wrong seventy-two years later.

Progress in Asia

For the one thousand years between 600 and 1600, neither Japan nor China made much progress in the development of pi. At the turn of the seventeenth century, the prevailing thinking in both countries put pi squarely at $\sqrt{10}$ (3.162 . . .), which was sometimes called the "root of perfection." The value was perhaps more idealistic than realistic, for it was significantly less accurate than the value found by Tsu Ch'ung-chih and his son twelve hundred years earlier (355/113, or about 3.1415929).

However, in 1663, Muramatsu Shigekiyo published his Sanso in Japan, demonstrating how to use an inscribed polygon to approximate the circumference of a circle. The Sanso was significant for two reasons. First, Muramatsu took the unusual step of showing how his result was accomplished. In Western history, this was the norm; however, in Japan, it was somewhat radical because the solutions to most mathe-

1 2 3 j A

matical problems were kept secret and were passed down only within distinct schools of math. Second, even though Muramatsu had calculated pi accurately to 7 digits, he wasn't fully confident of his findings and announced only that pi was approximately 3.14.

Muramatsu's method of polygons, like Archimedes', spread quickly and was used extensively throughout Japan for the next one hundred years. Seki Kowa found 16 accurate digits in 1712, and Tatebe Kenko calculated 40 digits by 1722. Also in 1722, Kamata Toshikiyo found and demonstrated another proof for pi, using both inscribed and circumscribed polygons (some Japanese writers have called this "squeezing the number pi from the inside and outside"), and calculated pi to the 24th decimal digit.

The proof was clear: Pi was just over 3.14. But curiously enough, most Japanese intellectuals stubbornly refused to listen to the newfangled proofs of lesser-known mathematicians who were bucking the system, and they continued to use a value of $\sqrt{10}$ until the nineteenth century and the end of the shogunate.

In China, a different story played itself out. After the very precise approximations of Tsu Ch'ung-chih, there followed a host of inaccurate values. As late as the mid-seventeenth century, Ch'en Chin-mo wrote that pi was exactly 3.15025 (his means are unknown). Two hundred years later, K'u Ch'ang-fa wrote the Wei-ching Chou-shu (The true considerations of circle measurement) in which he argued that the accurate value of pi was 3.125.

However, near the beginning of the eighteenth century, Jesuit missionaries such as Adam Schaal (better known by his Chinese name T'ang Jo-wang) and Pierre Jartoux (known as Tu Te-mei) began introducing Western scientific and mathematical thought to the Chinese. It's unclear how the Chinese were calculating pi before, but the missionaries may have also provided the Chinese with the key to measuring polygons. In

An equation by Tu Te-mei:

$$\pi = 3 \times \left[1 + \frac{(1^2)}{4 \times 6} + \frac{(1^2 \times 3^2)}{4 \times 6 \times 8 \times 10} + \frac{(1^2 \times 3^2 \times 5^2)}{4 \times 6 \times 8 \times 10 \times 12 \times 14} \right]$$

1713, a book entitled Su-li Ching-yun was published by imperial order. This book contained a chapter on calculating the ratio of the circumference to the diameter using inscribed and circumscribed polygons (beginning with a hexagon). The result of the calculations was 19 accurate digits of pi.

A hundred years later, in the early 1800s, Chu-Hung used Tu Te-mei's formula to find pi to 40 digits (of which 25 digits were correct). Later that century, the digit hunting took another leap when Tseng Chi-hung published his Studies in the Value of Circular Circumference. Tseng was the younger son of the famous statesman Tseng Kuo-fan, who led the imperial army during the Taiping Rebellion. Apparently, the younger Tseng had a greater fondness for math than for war, and by 1874, he had found 100 digits of pi using Gregory's series and the formula $\pi/4 = \arctan(1/2) + \arctan(1/3)$.

The Age of Electronic Computers: 1900 C.E. – the Present

The mysterious and wonderful pi is reduced to a gargle that helps computing machines clear their throats.

—**Philip J. Davis,** *The Lore of Large Numbers*

One hundred years ago, mathematicians only dreamed of using electronic machines to produce calculations as fast as lightning. However, by the middle of the twentieth century that dream became reality, and the electricity that moves lightning through the air began to be used to turn small switches on and off in binary sequences that could add and subtract numbers.

Remember that it took William Shanks many years of hand-cramping calculations to find his 707 digits of pi by 1873, and the accomplishment was hailed throughout the civilized world as the unveiling of a great mathematical truth. Lucky for him, the <u>real</u> truth—that he had made a mistake in his 527th place, and his last 180 digits were incorrect—was not found until long after his death, when, in 1945, D. F. Ferguson put pen to paper and found 530 correct digits of pi using an arctangent formula. It took him one full year to accomplish this—an average rate of just over 1 digit per day.

Fortunately, Ferguson was soon able to replace his pen and paper with the mechanical gears of an early desk calculator, and by September 1947, he had found 808 digits of pi. Desk calculators in the 1940s were about as similar to our modern-day slim-as-a-credit-card calculators as *Homo erectus* was to us, but they certainly added, subtracted, multiplied, and divided faster than people could. And with the threat of long-hand calculations removed, Levi Smith and John Wrench assumed responsibility for verifying Ferguson's work using Machin's formula—though it still took many months before the 808 digits could be completely confirmed.

When William Shanks first published his 707 digits in 1873, the errors weren't only in his calculations. Even before the infamous 527th digit, there were typographical errors introduced by the typesetter at the 326th digit, and other similar errors made at digits 460–462 and 513–515. In many ways, the value that Shanks first published in his 1853 treatise, *Contributions to Mathematics, Comprising Chiefly the Rectification of the Circle to 607 Places of Decimals*, was truly his most accurate work.

With that many digits in the bag, Smith and Wrench couldn't resist the temptation to keep going, and they found the 1,000th digit of pi in the winter of 1948. Nonetheless, the gears and keys of a mechanical desk calculator were clunky, and each step in figuring the necessary arctangents was laborious. At just 1 or 2 digits per day, it was hardly worth the effort.

D. F. Ferguson was at the Royal Naval College in England when he used the formula $\pi/4 = 3\arctan(1/4) + \arctan(1/20) + \arctan(1/1985)$ to find 808 digits of pi using a mechanical calculator.

Then, in 1949, the history of mathematics and the search for pi took a dramatic leap forward. This time, the breakthrough was not one of mathematical understanding, like

If you calculate the Gregory-Leibniz series to 500,000 terms, you'll get 30 places of pi. Unfortunately, it won't be entirely correct. Rather, you'll get: $3.14159\underline{0}6535897932\underline{40}4626433832\underline{6}$ . . . , where only the underlined digits are wrong.

the calculus; rather, it was the speed of the calculator itself. The American-made ENIAC (Electronic Numerical Integrator and Computer) was finally functional at the Aberdeen Proving Ground's Ballistic Research Laboratories in 1948, and the following year George Reitwiesner, John von Neumann, and N. C. Metropolis used this behemoth, with 19,000 vacuum tubes and hundreds of thousands of resistors and capacitors, to calculate 2,037 digits of pi. This calculation took merely seventy hours—including the time to handle the punch cards that had to be fed into the machine—averaging a digit every <u>two minutes</u>!

> Seven hundred seven,
> Shanks did state,
> Digits of π he would calculate
> And none can deny
> It was a good try
> But he erred in five
> twenty-eight.
>
> — Nicholas J. Rose

With the advent of the electronic computer, there was no stopping the pi busters, and only five years after the ENIAC calculated pi, computers had advanced to the point that the NORC (Naval Ordnance Research Calculator) was able to calculate 3,089 digits of pi in only thirteen minutes (about 4 digits per second).

Three years later, in 1958, the Paris Data Processing Center calculated the first 707 digits of pi in forty seconds on an IBM 704. Remember that ten years earlier Ferguson had taken over a year to complete this same feat, and a century earlier, Shanks had devoted a significant portion of his life to achieve it (albeit inaccurately). The scientists in Paris went on to calculate over 10,000 digits in one hour and forty minutes using Machin's formula and the Gregory-Leibniz series.

Then, only three years later, John Wrench and Daniel Shanks (not related to the earlier William Shanks) used an IBM 7090 to find 100,265 digits of pi, aver-

Even though Shanks and Wrench had figured more than 808 digits of pi by 1947, they purposefully limited their published value to that number in order to equal the 808-digit precision that P. Pedersen had found in calculating a value for <u>e</u>.

aging 3 digits per second—not including an additional forty-two minutes spent simply converting the result from binary to decimal digits.

Shanks and Wrench became famous for breaking the 100,000th decimal of pi on an IBM 7090 at the IBM Data Processing Center in New York. They used an equation found by Størmer in 1896:

$$\pi = 24 \arctan(1/8) + 8 \arctan(1/57) + 4 \arctan(1/239)$$

Each year, as computers would get faster, mathematicians would calculate pi to test the machine's speed and accuracy. In a race of "my computer is bigger than your computer," French, English, and American computer scientists vied for pi records throughout the 1960s and early 1970s in much the same way that the mathematicians of the seventeenth and eighteenth centuries did. The race climaxed in 1973 when Jean Guilloud and M. Bouyer found the 1 millionth digit of pi. (You, too, can find this digit; see page 130.)

Nonetheless, even though computers were getting faster and faster, the basic techniques for calculating pi weren't evolving as rapidly. Although the methods had improved over the years, brute force was the call of the day. But history has a way of repeating itself, and just as Snell and Wallis each broke away from the pack of polygon-based calculators in the seventeenth century, it was not long before someone had another stroke of brilliance that pushed the study of pi beyond mathematicians' wildest dreams.

In 1976, Eugene Salamin published an article in the Mathematics of Computation that demonstrated a quadratically converging algorithm for calculating pi. That is, the number of significant digits doubles after each step of the calculation, compared to earlier formulas which might provide 1, 2, or a small handful of additional digits per calculation. Later, Salamin found that his equation was similar to an algorithm discovered by Carl Friedrich Gauss over a century earlier for the eval-

Curiously, Australian Richard Brent independently developed a similar equation to Salamin's at almost exactly the same time.

uation of elliptic integrals. However, Gauss knew that it would be much too calculation-intensive to use this sort of equation to compute pi. Yet Salamin had something that Gauss did not—a computer that could handle millions of calculations per second. What was ludicrous to attempt in Gauss's time was relatively easy in the 1970s.

The combination of high-powered computers and the Gauss-Brent-Salamin algorithm rocketed pi calculations into the stratosphere. In 1982, Yoshiaki Tamura of the International Latitude Observatory at Mizusawa, Japan, and Yasumasa Kanada of the University of Tokyo calculated pi to 8,388,608 (2^{23}) digits. Using Salamin's algorithm on a Hitac M-280H, the calculation took just under seven hours.

Over the next ten years, two sets of brothers—Jonathan and Peter Borwein, and David and Gregory Chudnovsky—followed in Salamin's footsteps, developing high-powered pi algorithms. Their methods enabled mathematicians to calculate pi even faster and further. In fact, between 1988 and 1995, the Chudnovskys' and Dr. Kanada's computations leapfrogged each other. First, Kanada found over 100 million digits of pi, then the Chudnovskys found 525 million. And when Kanada found 536 million that same year, the Chudnovskys were stirred to reach 1 billion digits.

As we've seen before, knowing increasingly more digits of pi is hardly useful for any tangible application other than breaking in a new computer. But knowing more about the nature of pi may end up being important in our understanding of physics, geometry, and mathematics. So while the twentieth century has seen both our quantitative and qualitative understanding of pi skyrocket, the true richness of the number may not be found for many more years.

As this book goes to press, Kanada and Takahashi have calculated and verified just over 51 billion decimal digits of pi, setting a new world record. When you calculate a record number of digits, you can't be sure you've got the correct digits until you verify them. This generally means running the numbers a second time using a different algorithm. If the two values don't match, then a software or hardware bug must have crept in somewhere.

Since there are 360 degrees in a circle and pi is intimately connected with the circle, we eagerly look at [the 360th digit]. Again we are rewarded with a most remarkable fact. At [the 359th digit] we find 360. Thus 360 is "centered" over [the 360th digit]. —Monte Zerger, "The Magic of Pi," 1979

THE KID WHO LEARNED ABOUT MATH
ON THE STREET
If you divide 6,973 by 0, you die.
Once, this guy tried to find the square root of -9, and his eyeballs turned black.
This girl my brother knows found out exactly what π equals, but she went nuts.
R. Chast

Computing pi is the ultimate stress test for a computer—a kind of digital cardiogram. —Ivars Peterson, Islands of Truth, 1990

What is pi?

Mathematician: Pi is the number expressing the relationship between the circumference of a circle and its diameter.

Physicist: Pi is 3.1415927 plus or minus 0.000000005.

Engineer: Pi is about 3.

The universe was made on purpose, the circle said. In whatever galaxy you happen to find yourself, you take the circumference of a circle, divide it by its diameter, measure closely enough, and uncover a miracle—another circle, drawn kilometers downstream of the decimal point. . . . As long as you live in this universe, and have a modest talent for mathematics, sooner or later you'll find it. —Carl Sagan, Contact

If a string were tied around the equator of the earth (assuming the earth were perfectly round, which it isn't), the string would have to be 2π feet longer in order to sit 1 foot off the surface of the globe.

I knew—but anyone could have sensed it in the magic of that serene breathing—that the period was governed by the square root of the length of the wire and by π, that number which, however irrational to sublunar minds, through a higher rationality binds the circumference and diameter of all possible circles. The time it took the sphere to swing from end to end was determined by an arcane conspiracy between the most timeless of measures: the singularity of the point of suspension, the duality of the plane's dimensions, the triadic beginning of π, the secret quadratic nature of the root, and the unnumbered perfection of the circle itself. —Umberto Eco, Foucault's Pendulum

A Pi Chronology

c. 2000 B.C.E.	Babylonians use $\pi=3\,^1/_8$; Egyptians use $\pi=(256/81)=3.1605$.
c. 1100 B.C.E.	Chinese use $\pi=3$.
c. 550 B.C.E.	Old Testament implies $\pi=3$.
c. 434 B.C.E.	Anaxagoras attempts to square the circle.
c. 430 B.C.E.	Antiphon and Bryson articulate the principle of exhaustion.
c. 335 B.C.E.	Dinostratos uses the quadratrix to "square the circle."
third century B.C.E.	Archimedes uses a 96-sided polygon to establish $3\,^{10}/_{71} < \pi < 3\,^1/_7$. He also uses a spiral to square the circle.
second century C.E.	Claudius Ptolemy uses $\pi = 3^\circ\, 8'\, 30'' = 377/120 = 3.14166\ldots$
third century C.E.	Wang Fau uses $\pi = 142/45 = 3.1555\ldots$
263 C.E.	Liu Hui uses $\pi = 157/50 = 3.14$.
c. 450	Tsu Ch'ung-chih establishes 355/113.
c. 530	Aryabhata uses $\pi = 62{,}832/20{,}000 = 3.1416$.
c. 650	Brahmagupta uses $\pi = \sqrt{10} = 3.162\ldots$
1220	Leonardo de Pisa (Fibonacci) finds $\pi = 3.141818\ldots$
1593	François Viète finds first infinite product to describe pi; Adriaen Romanus finds pi to 15 decimal places.

1596	Ludolph Van Ceulen calculates pi to 32 places.
1610	Van Ceulen expands calculation to 35 decimal places.
1621	Willebrod Snell refines the Archimedean method.
1654	Huygens proves the validity of Snell's refinement.
1655	John Wallis finds an infinite rational product for pi; Brouncker converts it to a continued fraction.
1663	Muramatsu Shigekiyo finds seven accurate digits in Japan.
1665–66	Isaac Newton discovers calculus and calculates pi to at least 16 decimal places; not published until 1737 (posthumously).
1671	James Gregory discovers the arctangent series.
1674	Gottfried Wilhelm Leibniz discovers the arctangent series for pi.
1699	Abraham Sharp calculates pi to 72 decimal places.
1706	John Machin calculates pi to 100 places; William Jones uses the symbol π to describe the circle ratio.
1713	Chinese court publishes Su-li Ching-yun, which shows pi to 19 digits.
1719	Thomas Fantet de Lagny calculates pi to 127 places.
1722	Takebe Kenko finds 40 digits in Japan.
1748	Leonhard Euler publishes the Introductio in analysin infinitorum, containing Euler's theorem and many series for π and π^2.

1755	Euler derives a very rapidly converging arctangent series.
1761	Johann Heinrich Lambert proves the irrationality of pi.
1775	Euler suggests that pi is transcendental.
1794	Georg Vega calculates pi to 140 decimal places; A. M. Legendre proves the irrationality of π and π^2.
1844	L. K. Schulz von Stassnitzky and Johann Dase calculate pi to 200 places in under two months.
1855	Richter calculates pi to 500 decimal places.
1873	Charles Hermite proves the transcendence of e.
1873–74	William Shanks publishes his calculation of pi to 707 decimal places.
1874	Tseng Chi-hung finds 100 digits in China.
1882	Ferdinand von Lindemann proves the transcendence of pi.
1945	D. F. Ferguson finds Shanks's calculation wrong from the 527th place onward.
1947	Ferguson calculates 808 places using a desk calculator, a feat that took about one year.
1949	ENIAC computes 2,037 decimals in seventy hours.
1955	NORC computes 3,089 decimals in thirteen minutes.
1959	IBM 704 (Paris) computes 16,167 decimal places.

1961	Daniel Shanks and John Wrench use IBM 7090 (New York) to compute 100,200 decimal places in 8.72 hours.
1966	IBM 7030 (Paris) computes 250,000 decimal places.
1967	CDC 6600 (Paris) computes 500,000 decimal places.
1973	Jean Guilloud and M. Bouyer use a CDC 7600 (Paris) to compute 1 million decimal places in 23.3 hours.
1983	Y. Tamura and Y. Kanada use a HITAC M-280H to compute 16 million digits in under thirty hours.
1988	Kanada computes 201,326,000 digits on a Hitachi S-820 in six hours.
1989	Chudnovsky brothers find 480 million digits; Kanada calculates 536 million digits; Chudnovskys calculate 1 billion digits.
1995	Kanada computes 6 billion digits.
1996	Chudnovsky brothers compute over 8 billion digits.
1997	Kanada and Takahashi calculated 51.5 billion (3×2^{34}) digits on a Hitachi SR2201 in just over 29 hours.

This table was compiled from a number of sources, including Petr Beckmann's A History of π.

In 1777, naturalist George Louis Leclerc, comte de Buffon, posed the following question: If you throw a needle of length L onto a table marked with evenly spaced parallel rules (with d distance between them), what is the probability that the needle will cross one of the lines? As long as d is larger than L, the answer is $P = 2L/\pi d$. If you toss the needle enough times, you can actually "calculate" pi—by dividing the number of hits by the number of tosses (P) and then solving for pi ($\pi = 2L/Pd$). This is often called the Monte Carlo method. Of course, this is a pretty poor method; even if you throw the needle thousands of times, you're unlikely to get more than 3 or 4 digits of pi.

$$\sqrt{\frac{1}{2}\pi e} = 1 + \frac{1}{1\times 3} + \frac{1}{1\times 3\times 5} + \frac{1}{1\times 3\times 5\times 7} + \cdots + \cfrac{1}{1+\cfrac{1}{1+\cfrac{2}{1+\cfrac{3}{1+\cfrac{4}{1+\cdots}}}}}$$

Ramanujan's Insight

Srinivasa Ramanujan was born in 1887 in a small town in southern India to a family without wealth. He grew up learning math alongside his schoolmates, but it quickly became clear that he was a prodigy in the field. In fact, he enjoyed math so much that his other studies suffered, and his college education ended when he failed his exams in other required subjects.

By the time he was twenty-five, Ramanujan was married and working as a low-paid clerk in Madras. Even then, he did not stop experimenting with mathematics, writing equations prolifically in his notebooks. His equations were quite varied, including many solutions for approximating pi, and he rarely demonstrated a proof or showed his methods. He was alone in his world of numbers, calculating and theorizing without peers who could understand, much less advise.

Then, in 1913, he sent several pages of his discoveries to three prominent English mathematicians. Two of them rejected his work, probably thinking him some sort of crackpot. When the third mathematician, G. H. Hardy, received Ramanujan's letter, he decided that "they [the discoveries] must be true, because if they were not true, no one would have had the imagination to invent them."

The next year, Hardy brought Ramanujan to England to collaborate on a number of mathematical projects. The two often traded roles of teacher and student. Although Ramanujan had no classical training in mathematics, his intuitive understanding of the subject was phenomenal. Ramanujan learned not only the ways of classical mathematics but also those of Western culture. He found eating with utensils and binding his feet into shoes to be most distasteful, not to mention the acute difference in climate between tropical southern India and his cold, damp new home.

Soon after World War I began in 1914, Ramanujan became chronically ill and was in and out of sanitoriums for several years. It's likely that his strict Hindu vegetarian diet, almost impossible to maintain in war-torn Britain, caused a severe vitamin deficiency. By 1919, when passage back to India was once again safe, he returned home. And though his sickness continued and he was wracked with pain, Ramanujan kept filling his notebooks with equations, once again rarely showing how he obtained his results. He died a year later, at the age of thirty-two.

Seventy years later, scientists and mathematicians are still unraveling the fascinating equations of this genius, applying them to current-day problems and using them to generate other algorithms that are designed to be efficient when run on computers. In the mid-1980s, Jonathan and Peter Borwein developed an extremely powerful equation to calculate pi based on studies of Ramanujan's formulas.

These are iterative equations—you can plug the result of the calculation back into the formula to get an even closer approximation to pi. The results are incredible because you can double or even quadruple the number of relevant digits each time you run the formula.

$$\frac{1}{\pi} = \frac{2\sqrt{2}}{9801} \sum_{n=0}^{\infty} \frac{(4n)!}{(n!)^4} \times \frac{[1103 + 26390n]}{(4 \times 99)^{4n}}$$

Ramanujan, like most mathematicians, could not resist the temptation to explore pi, and his intuitive leaps brought amazing insights to the study of the number. What might he have found had he lived longer? The answer to that question is as shrouded in mystery as pi itself.

$$\pi \approx \left(9^2 + \frac{19^2}{22}\right)^{1/4} = 3.14159265262$$

$$\int_0^\infty \frac{(\log x)^2}{1+x^2}\,dx = \frac{\pi^3}{8}$$

24138121771320340323718671520627359281633744103154710463963111770834535015448259457808665977571739145766095877536672498405172457008529582124548521961643789313109882481700384223320
632885004325420849215944237347069301246933391888763956241425406832442961396568624031641284030587825646523428183365665189585503699094329530942506080824684142553143703739066510989
6765398087358749937703345895301949519517021752264520732035027546394131137116116222899004570802847540361406381477089041361896398146793609058294820541858067653745684521520114514513358
47400424917141834979108526755784972364608405102964056854716291147960516605129860116421892338470677447838697916343692200302198889638490152417264835108736731345893534421520021661553366
3361483478642335613805300749667620842875530742848476251219552064026842561574021989849634090361092184760431288829692663808182477416243214918051745760516527671485956723605854814854814
544021329075323119337897054213766925989414624437580625161532179973469637179645547335354924900140560743663106471766798573256986302528399444346379930518242598412579835864959562789069
69536518549591816028252311296488624586662474149878023489619904934728196665830714757725201586835547077836728257562399243554639937945789912258525500166665977501754113554118015591419
39804521789111868368461117336749565952373215881922246966429594969400736850502840377689062658085002465172958976399152885528712794692079212206732032052809396452263227086822123147520
80066037858618410693456045525790977739556180123341874179413621578605007839034591252524540485754363270989363738481374250658388020484570153291728775365851028430430378094645827944632
31703476961344868280054938747420783607209181969560779780786595574079606443029786093090797207307496070318155868501394897534530052729341161714831473396610051752170023014799010869
0345299770487759840572862989418102152969007455565972433483618503777150550860330115053176164169618772366138127287871098651734520385737873021661722258062394563423895118271063899993
82819394680908971742686787488703236974236982543558883246058458520840234832623588364349326648017599040342821517219563963495304304081884166217827515149492540901197488825950004047265
569119457920367936540376113867493761913528838976548861213181153010471606196867043644824884349888225522312968750291635458654286806894604692937059649512844489
74048576700358527040399356260489828221525557196999525794545216174074323708599113001429067140022594327460920190862992179553442748718116421174293434646185812595775017541335441801559014
001252399489390176617521619230063203926935030744080055648532173648116815833020317058976467823292022381824767644909239971574846690066268487079269797454361050026265979711717212565458
187223595567184718953968963563982739116945400828677220839735648525019605960672645515193429252306818637594677165747716398510357
58010264645131490589465320117025939029809262215532611811216870595841647293227196915373233515149878813034739889849652542170463108520020303089349760747809669523143814489522362876939315704764290005257909875253160574092449263176141128160500604332678149217024478104090000252357290745094008754190944850132342773779280281036876877759883889102242894658630718857836325840114004029582015572777505520049767171741806532996812614435963077468399197436150775560849014830448152622616887549680346283040684967488458514489661089841116418534724670949542984233687929502850553562738086930362906343961063890273423981644437129896752462213499807179098325383537518245143182298170149804714744052082067717348290375742446077778472524596418574890493050596507992590542253069212603701703789185714698302257724852517384591456079105588537047066293052686115149625701677560509871522189311993190860804059893272714652003150911804363740649554498522116311079240253084120858743275604673359050242874200960582178254070725358194242742829061261156506709060900696602286669193502613356980448499030606097708791796420344947066473435831304989532397059589076521205958976976179995460990255750129529505173644332819377848179827289210626883079150390928415489284480545010183293430169393085976918826070982327288892113515169823445644473332530729629985779235645767408449474154678075407907903336969675987555609854817554818183668884928926249289370935702416477065288458257586878431245022897907054067823608079140471148112227376233667965016689209723738120832311982620373778281816414460207482821822157164029456589716075407686789175255031704962919619171215072124527331363754728630499702245934560313114447968406216583475010344929717322069459738127450279457465466369080110231031103053073805061028776263848004279278979597037886652702744102612606837900772640917150377133336149971740090051602135428701865990605371259092989988875726878006791586909108457040780273990101875834025072067523229798455564584723383879369483932121937056631027358110963094423462935735458743954610171509748417603259483536217516712490048287878693443178634077789561343147653044727101587308359186

There's a beauty to pi that keeps us looking at it. . . . The digits of pi are extremely random. They really have no pattern, and in mathematics that's really the same as saying they have every pattern.
—Peter Borwein, 1996

To ask for the system in pi is like asking "Is there life after death?" When you die, you'll find out.
—Richard Preston, "The Mountains of Pi," The New Yorker

The circle is one of the noblest representations of Deity, in his noble works of human nature. It bounds, determines, governs, and dictates space, bounds latitude and longitude, refers to the sun, moon, and all the planets, in direction, brings to the mind thoughts of eternity, and concentrates the mind to imagine for itself the distance and space it comprehends. It rectifies all boundaries; it is the key to information of the knowledge of God. —John Davis, The Measure of the Circle, 1854

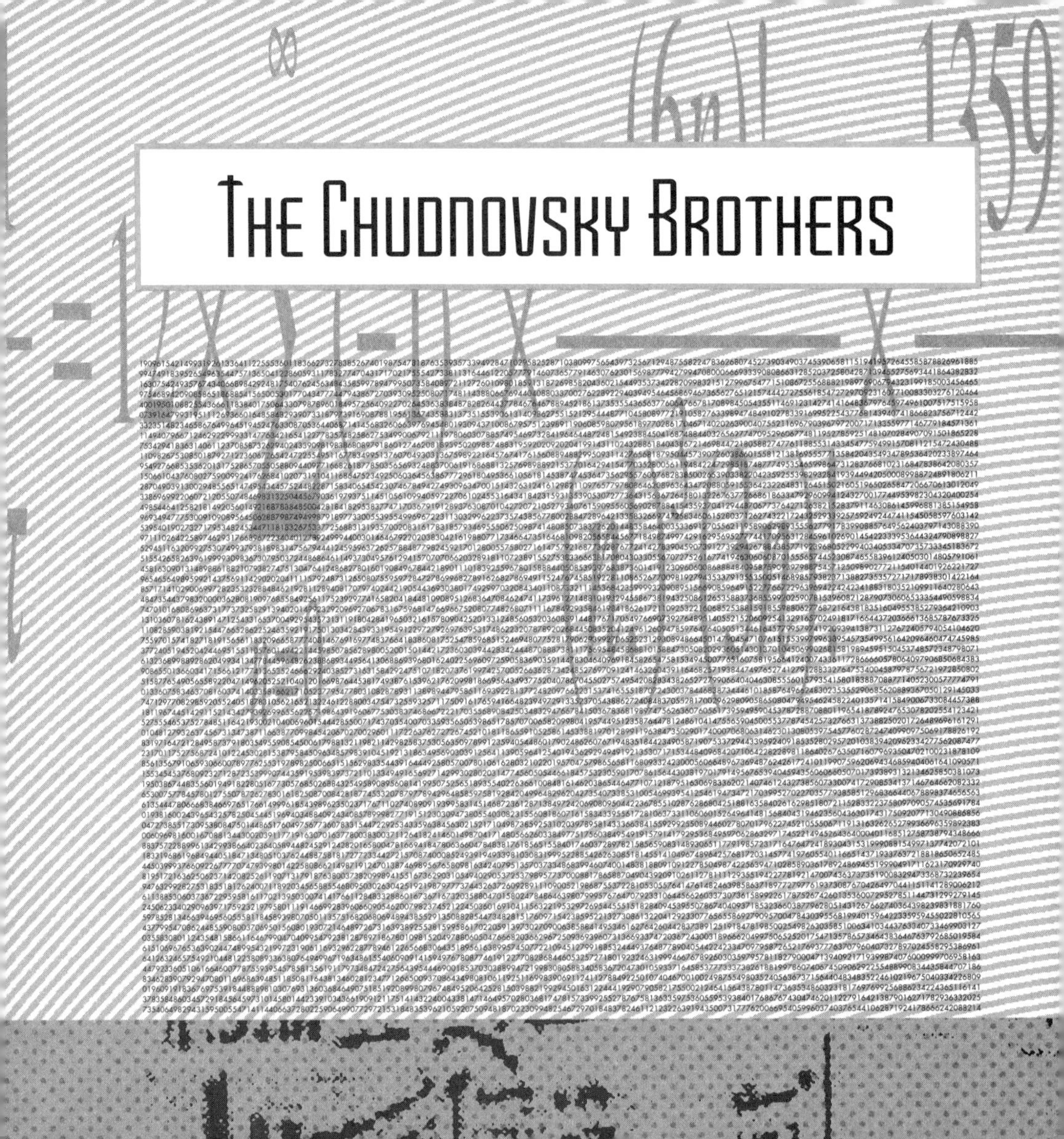

The Chudnovsky Brothers

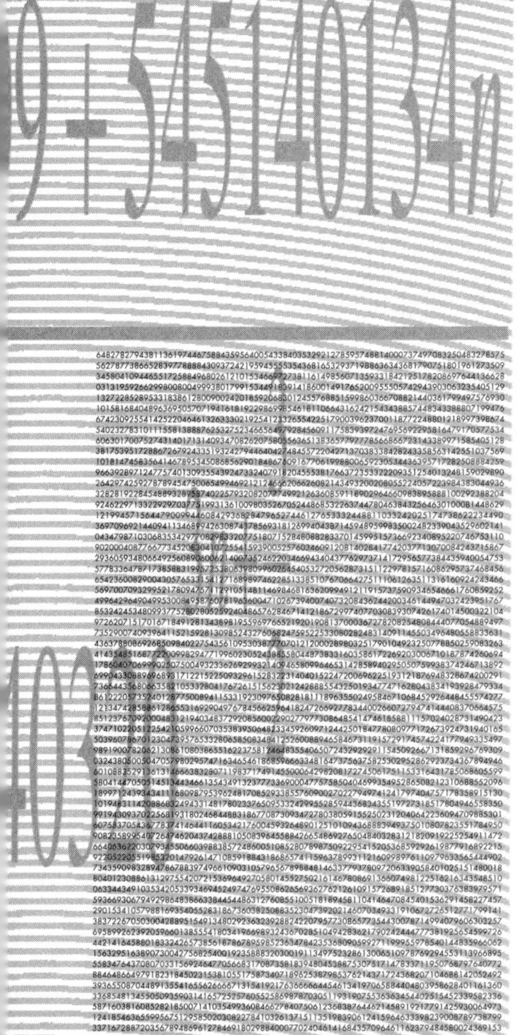

Although the decimal expansion of $\pi/4$ is not literally random, it is as random as one would need for most purposes: it is "pseudorandom." Mathematics is full of pseudorandomness—plenty enough to supply all would-be creators for all time.

—**Douglas Hofstadter,**
Gödel, Escher, Bach

In every generation, a handful of people pursue pi to its extreme, either by calculating more digits or by delving into a deeper understanding of the number through equations and formulas. David and Gregory Chudnovsky are two mathematicians who are doing both. The brothers have held several world records for calculating the highest number of digits (first at 450 million digits, then 1 billion, and then 2 billion), and they have also developed extremely sophisticated equations to describe pi.

Both world-class number theorists, the

Chudnovskys want to understand this fascinating transcendental number better, to learn about its nature and personality by immersing themselves in the digits. But these mathematicians are not white-coated members of a well-funded laboratory, using supercomputers plated with exquisitely burnished steel and designer-black walls. They are not well-paid visiting lecturers at Ivy League colleges.

Rather, David and Gregory Chudnovsky are fascinating characters in the story of pi because they are brilliant hackers who, in their mid-forties, built a supercomputer in their apartment from off-the-shelf parts in order to harvest transcendental numbers and explore infinite series at their leisure.

The Chudnovskys were born in the former Soviet Union, near Kiev, in the years following World War II. Their parents, who were scientists and intellectuals, introduced them to mathematics at an early age. While both brothers enjoyed math, it became clear that Gregory, five years younger than David, was a brilliant student of the field. Encouraged by his family, he published his first mathematical paper at the age of sixteen. Later, both brothers studied mathematics in college and received Ph.D.s from the Ukrainian Academy of Sciences.

However, Gregory, diagnosed with myasthenia gravis, an autoimmune disorder of the muscles, soon became weakened to the point of being frequently bedridden. By

1976, the illness became bad enough for their parents to attempt to leave the Soviet Union in order to seek treatment for Gregory. The Soviet government was, predictably, less than helpful, and the Chudnovskys withstood a barrage of harassment from government officials. At one point David and his parents were physically beaten by KGB agents. Fortunately, word of Gregory's predicament spread, first through the world's mathematical community and then into the political arena. With global pressure mounting, the family was permitted to leave in 1977, and they immediately emigrated, first to Paris, and then to the United States.

Until recently, Gregory's apartment on the Upper West Side of Manhattan was the brothers' computational battlefield. In the late 1980s, before they built their own supercomputer, they calculated pi remotely on two supercomputers (a Cray 2 at the Minnesota Supercomputer Center and an IBM 3090-VF at the IBM Thomas J. Watson Research Center in New York). That's not unusual, but unlike most scientists, Gregory Chudnovsky had to write, run, and monitor their pi-calculating program while lying in his bed.

However, renting supercomputer time is not only extremely expensive but also frustrating if (or, more likely, when) anything goes wrong, for you have no control over the device itself. It was soon after their first record-breaking pi experience (480 million digits in 1989) that the brothers began to build their own computer, dubbed m zero, in Gregory's apartment. m zero was a multiprocessor computer that could reach incredible speeds of many billions of operations per second. Of course, this was no small desktop model; the computer took up much of the apartment, with various components snaking from one room to another. Computers always generate some heat, but m zero used so much power that it could raise the ambient room temperature—even with air-conditioning—to above 90 degrees Fahrenheit.

Having your own supercomputer is a mixed blessing. While it can be helpful when

44623192970276370458105466861864892006481857884470461376294282061421070348036308441116780061481502391369013966575803093965904415912512096952973581273097903258329274793996928243831492062590293309286010332391
74038383427457113513984577478320244883860219680767163316534986730347454013999215982211167111512544258537609167343276999904005844974347590556420630769469347283565064991077767958926507983234810840182244891 1707
49665104061875918474857697210367432681514842019569722074734482208792869960836293587631653890478390048742747935151528419727078614321965828417069390496815772568280501220207510923634655176931515723238105018022
585517518247822341364317881652068506613942262653446972893597805755792607832166738898066125192388000169534325222620052889956108613287995075957355474955247424308328701918030564770827367014344539375937631550175
0935185158798505069463905231958292523097514074052545956314007436631365318598857577378878419026141118689544565042160337064786252627247085434175977950069264902695111202750263095436740580164721315926794537943694
97526101844660858000783127191316021235091382970425870233641604644846128470918634315198653654660902676182247471801225355996813235223866795559736925806895975303712648024482045813701820343286908637165983377571 0
85833010368743465774110621814317655633962100638583167467480918871707737653558986622029499873827444255479241163465467940603650247460245618952590934996673595122712732052401111139185230270676142772309222947288
10133192963339978185503182934432632206469801012951704557043006325015625043150836576283267412300204950639717367841889100514252530524968862567387384777966286621649736240288369309617354373373136677583583570179

attempting cutting-edge math, the Chudnovskys—who are number theorists—have had to learn the intimate workings of computer programming and computer hardware design. It's the difference between just knowing how to drive a car and designing your own souped-up roadster for the Indianapolis 500 and then racing it yourself.

So now, instead of thinking about numbers—their nature, their personality, their relationships—the brothers must also sit and make decisions about buying hard-drive controllers, reducing the temperature of logic boards, and how best to make archaic software languages run over parallel processors with the utmost efficiency.

And yet, it is a price they're willing to pay, and the results of their work are often spectacular. The brothers were once quoted as saying—though you often don't know which brother said what, since they frequently interrupt each other and finish each other's sentences—"We are looking for the appearance of some rules that will distinguish the digits of pi from other numbers. It's like studying writers by studying their use of grammar. If you see a Russian sentence that extends for a whole page, with hardly a comma, it is definitely Tolstoy. If someone gave you a million digits from somewhere in pi, could you tell it was from pi? We don't really look for patterns; we look for rules."

The digits of pi appear so random that if there were a rule to the sequence, it may require billions—or trillions—of digits to begin to see it. The Chudnovskys have said that no other calculated number comes closer to a random sequence of digits. "Pi is a damned good fake of a random number," Gregory once said. "I just wish it were not as good a fake. It would make our lives a lot easier."

The brothers appear to have been given few breaks that would make their lives easier. While

MATH MAJORS EXCHANGING HIGH πs

many of the world's leading mathematicians agree that Gregory is one of the most astounding minds of our time, in the twenty years that they've been in this country, neither Gregory nor David (who's a pretty spectacular mathematician himself) has been able to secure a long-term job. All kinds of excuses have been given, including Gregory's obvious physical disability and David's not-so-subtle Russian brashness. Nevertheless, mathematicians have noted that Gregory knows things about hypergeometric functions that no other living person understands, and when he eventually dies, the knowledge may well be lost.

> There are no occurrences of the sequence 123456 in the first million digits of pi. But of the eight 12345s, three are followed by another 5. The sequence 012345 occurs twice, and in both cases it is followed by another 5.

Fortunately, in 1981, Gregory was awarded a coveted MacArthur Foundation fellowship in mathematics. This so-called genius award offered not only a generous yearly stipend for several years but also full comprehensive medical coverage for him. The monies from this, plus additional sums culled from their wives' paychecks, have been used to build, maintain, and upgrade m zero over the years. It's a labor of love, but also one of infinite fascination with numbers.

The Chudnovskys are often like children with a new idea; not childish at all, but rather incredibly playful and creative. They can take a number and turn it into a game or a puzzle, and they can't help but sprinkle humor throughout their conversation. Gregory says that calculating pi is like participating in extreme sports. "It should not be tried at home," says David.

By the spring of 1989, the brothers took the world by surprise by doing just that: calculating 480 million digits from their home. Mathematicians had known of the Chudnovskys for years, of course, but they never knew that the brothers were interested in pi. To this day, many people think that calculating digits is a frivolous activity, and for such great minds to be preoccupied with it was curious indeed. On the other hand, perhaps the fact that people with such an intuitive sense of mathematics have been drawn to the circle ratio is itself evidence that understanding pi might draw us closer to a deeper comprehension of math and physics itself.

Still, there are always obstacles in the way. Calculating a record-breaking number of digits for pi pushes the envelope for any computer, and every tiny flaw in hardware or software bubbles to the surface in the attempt. Errors like those found in the Pentium processor a few years ago—where a mistake may occur only once in a billion calculations—are not uncommon when you push a supercomputer at over 100 billion calculations every minute for days, or even weeks, at a time. Finding those tiny fractures of logic, or miniscule oversights in a design, can add months of work to the process. But they must be found, for if any one digit in pi is calculated incorrectly, every subsequent digit is likely to be wrong as well.

> The sequence 3333333 appears at the 710,100th digit and again at the 3,204,765th digit. Not so surprising, really. In fact, the first million digits include seven-long runs of the same number for each numeral other than 2 and 4.

Pi is certainly not the only constant that Gregory and David Chudnovsky are interested in. They have worked informally with Columbia University's applied mathematics department for years, focusing on nonlinear dynamics of all sorts, and have written extensively on a number of topics.

And as this book goes to press, the brothers have just created a new office at Brooklyn Polytechnic University, called the Institute for Mathematics and Advanced Supercomputing, and are preparing to continue their work from there. David explains, "The institute has just two members, Gregory and me." It may be a perfect solution for the "Chudnovsky problem," because in this position they can perform their own work and yet still contribute to the university. Gregory may even finally be able to teach from home, via video teleconferencing.

The new institute was also an impetus for the brothers and their wives to move out of Manhattan; unfortunately, many pieces of m zero had to be disconnected and packed into boxes. Just a few months before tearing it down, they calculated pi one last time in order to test some upgrades they had recently made. After a week of elapsed computing time, the brothers had a new world's record: over 8 billion digits calculated and verified. At this point, they didn't much care about the record, but they were happy their computer and their algorithms were working correctly. "This is not mathematical trickery," David insisted. "This is very serious mathematics."

But the brothers' record-breaking calculations and exquisite formulas—developed as much from an extraordinary understanding of math as from an unflagging dedication to their cause and their homespun computer—have earned the Chudnovsky brothers a permanent place in the history of pi.

$$\frac{1}{\pi} = 12 \times \sum_{n=0}^{\infty} (-1)^n \times \frac{(6n)!}{(n!)^3(3n)!} \times \frac{13591409 + 545140134n}{640320^{3n+\frac{3}{2}}}$$

Think of games for children. If I give you the sequence one, two, three, four, five, can you tell me what the next digit is? Even a child can do it; the next digit is six. How about this game? Three, one, four, one, five, nine. Just by looking at that sequence, can you tell me the next digit? —Richard Preston, "The Mountains of Pi," The New Yorker

Exploring pi is like exploring the universe. —David Chudnovsky

It's more like exploring underwater. You're in the mud, and everything looks the same. —Gregory Chudnovsky

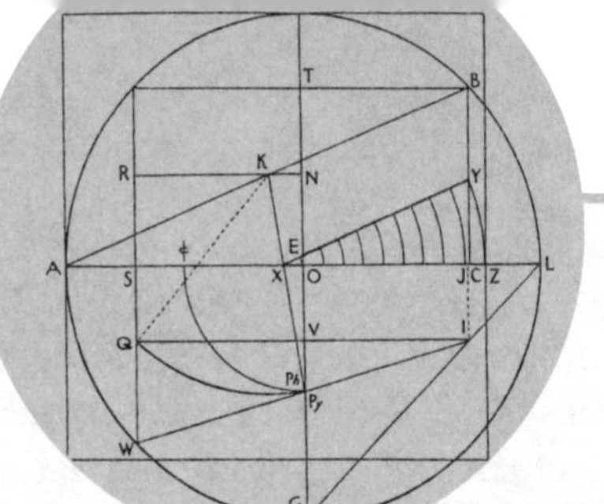

What's Inside π?

It's not poetic or particularly pleasing to hear, but we humans are basically pattern recognition devices. Our eyes take in the world, but what we really see are intricate patterns of lines and curves and colors and brightness. Our ears hear sound, but we only recognize music and language when we decode the signals into discreet patterns of tone and rhythm.

We crave finding patterns in the world around us because it's the only way that we can give meaning to anything, including ourselves. Nature may loathe a vacuum, but humans cannot stand a lack of pattern. We're programmed to find patterns around us, and we'll stop at nothing to find them—even in pi.

The digits of pi reel off into infinity in a stream that appears to be entirely random. Like rolling a 10-sided die, there's a 1-in-10 chance that whichever number comes next in line will be 3, or 7, or 0. But the digits of pi aren't just meaningless cosmic noise; if you change a single digit—perhaps you replace a 4 with a 9—you no longer have pi. The ratio of a circle's circumference to its diameter demands precision.

It's partly this paradox of the simplicity of a circle and the apparent randomness of the digits of pi that has driven people to hunt down billions of digits of the number. We're looking for the pattern we know "should" be there, and the longer we stare at the digits, the more they, too, appear to dance and cluster before our eyes, like static on a television screen.

If you correlate the digits of pi to notes on a musical scale, there is no melody, but listen to it long enough and you start to think you hear one. Assign colors to the

In his article entitled "The Magic of Pi," Monte Zerger pointed out that the 7th, the 22nd, the 113th, and the 355th digits of pi are all 2s (he counts the 3 before the decimal point as a digit). Remember that 22/7 and 355/113 are two of the best approximations for pi. The next best approximation for pi is 52163 and 16604. Curiously, digit number 52,163 is also a 2. Digit 16,604 is unfortunately a 1, though the digit before and after it is a 2.

digit, and you can faintly make out a . . . or perhaps it's not there after all.

In the nineteenth century, Augustus De Morgan dipped into the first 600 digits and noticed that there are fewer 7s than you would expect. (You'd think there would be sixty, but in fact there are ten too few.) What could it mean? Could circles naturally be 7-phobic? Of course not; it's just that 600 digits isn't enough of a sample to really get a good picture, statistically speaking. So, in our drive to find possible patterns in pi, we must have a lot of digits.

It wasn't until the 1950s that we could explore 10,000 digits of pi, at which point the distribution of the digits starts to even out. And now, at over 51 billion digits on record, it appears that there's no statistically relevant difference between the number of 7s or the number of zeros, or any other number.

And yet, we can see that there are 17,176 more 5s than would be expected. Maybe—just maybe—if we calculate out another few billion digits, we'll start to see a pattern that could help us get our heads around this elusive number and possibly, in turn, the universe around us.

The matter of the normalcy or nonnormalcy of pi will never, of course, be resolved by electronic computers. We have here an example of a theoretical problem which requires profound mathematical talent and cannot be solved by computations alone. The existence of such problems ought to furnish at least a partial antidote to the disease of computeritis, which seems so rampant today. —Howard Eves, Mathematical Circles Revisited, 1971

Mathematician John Conway pointed out that if you break down the digits of pi into blocks of ten, the probability that one of those blocks will contain ten distinct digits is about one in 40,000. Curiously, this first happens in the seventh block of ten digits.

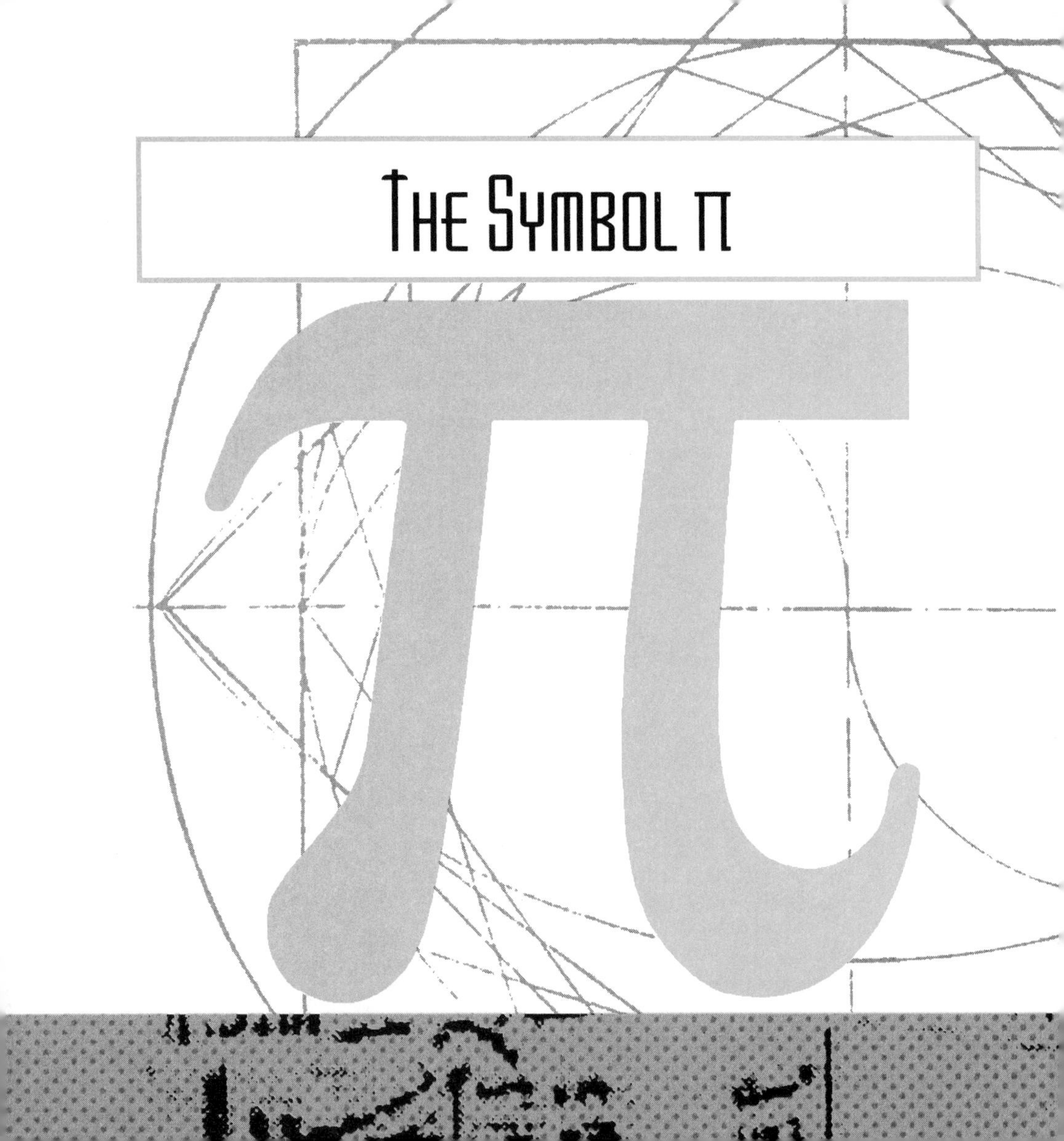

The Symbol π

If we take the world of geometrical relations, the thousandth decimal of pi sleeps there, though no one may ever try to compute it.

—**William James,**
The Meaning of Truth

π, the sixteenth letter of the Greek alphabet, is the most widely recognized and used Greek letter outside of the American university fraternity and sorority system. It appears on the keypads of calculators. It appears in textbooks on most of the sciences, including math, statistics, physics, and astronomy. It resonates for both elementary school students and learned university professors. Even when people can't remember exactly what π signifies, they recognize the symbol.

It's astounding, but the ancient Greeks themselves did not use the symbol π to signify the ratio of a circle's circumference to its diameter. Nor did the Romans, nor the Arabs, nor the Chinese. In fact, hardly anyone used *any* single symbol to denote this ratio until two thousand years after Archimedes studied circles. The symbol π has only been used regularly in its modern meaning for the past 250 years.

One of the earliest uses of the symbol π in mathematics was by William Oughtred in 1652. He defined the ratio of the circumference to the diameter as π/δ, where π clearly stood for *periphery* and δ for *diameter*. A few years later, John Wallis used □ (a box) or the Hebrew letter *mem* to signify what we'd call $4/\pi$. Later, in 1685, Wallis used π to represent the *periphery* described by the center of gravity of a body in a revolution.

The first time anyone used a single symbol to refer to the circle's ratio, that symbol was certainly not π. In fact, in 1689, J. Christoph Sturm, a professor at the University of Aldorf in Bavaria, used the letter *e*. Suffice it to say that this notation did *not* catch on. Over the next one hundred years, most other mathematicians continued using a ratio of two symbols—such as π/ρ or *c/r* (where both ρ and *r* stand for the radius of the circle).

You can determine your hat size by measuring the circumference of your head, then divide by pi, and round off to the nearest one-eighth inch.

In his Synopsis palmariorum matheseos, published in 1706, William Jones used π in its modern understanding for the first time. Math historian Florian Cajori writes about this historic step in his 1928 classic, A History of Mathematical Notations:

> It was in that year that William Jones made himself noted, without being aware that he was doing anything noteworthy, through his designation of the ratio of the length of the circle to its diameter by the letter π. He took this step without ostentation. No lengthy introduction prepares the reader for the bringing upon the stage of mathematical history this distinguished visitor from the field of Greek letters. It simply came, unheralded, in the following prosaic statement:
>
>> There are various other ways of finding the Lengths or Areas of particular Curve Lines, or Planes, which may very much facilitate the Practice; as for instance, in the Circle, the Diameter is to the Circumference as 1 to
>>
>> $$\overline{\frac{16}{5}-\frac{4}{239}}-\frac{1}{3}\,\overline{\frac{16}{5^3}-\frac{4}{239^3}}-\text{\&c.}=3.14159,\ \text{\&c.}=\pi.$$
>>
>> This series (among others for the same purpose, and drawn from the same principle) I received from the Excellent Analyst, and my much esteem'd Friend Mr. John Machin.

It is interesting to note that Jones had also used π in various other ways earlier in his book. For instance, in one place π was a point in a geometric figure; in another place, Jones used π to represent periphery. The vacillation of his terms was common for the time, and would continue to appear in the works of finer mathematicians than he.

738220743145689445864371754121066049337745424904700149039594657459437712378972800584293962105526750395663214923767298140663500612023607935950725002611882298817024409323060665400972298243353765243779941 59185
149339041722501469974253353780504362141093577235490088186907390269514016697214836261672332385111763897273755722811689919980399572937460322172819284534093325844116963040623830035426887429021182429856451 24579
520734798462734619510455242561373629943301498393729111087052996951918353365053484424284804631047652208342085936517309658449613028583365481954359818675622029416138432116325768598256296720182890678112418686 87
845973133871404187297007119198677012931062930969498993548139218759288048398451387407586430276561571473351835440735062553123436649600531091488130323844462652043513629026265933306812051158885104358569945649938
869570706111923572757110294196928994187655436884256928063861435130310638444334313396196979733561523242666641706639024036236559357494425103835682493854306636956562838134975783190902427704990944514111124123020
604342232889259749143823157590798442351778454112872425949353333098571097875538683981754825212175181030821817650515556345242994845390457756267446578551759148916225117807053288117045974059759864583008005 61032
233253843007509811343101204508331904228654685877994401229656165638908159592239403439222600101972217657362171163599025757529000338660227492594219355961294755185136993932496927866350652388267689411676389 08982
314619095968338612311858671495757270575686399745427113036591818384717808180841752010065566213047632675249466820599746393688064999076413425808930820409023632383517830632217617206433632011040344099409158 40514
687832626047756804071261548426034683388026809404479308919373423903638644025492448115739076316802664679967901755671870641363324028870508745716587139591642925361440259784029087134377444175895665581130737 62968
893475271737113137790003100827293871224879869141242800845027251225463527219920187624235080278447967843370236807361439928590481111125419311013509593127276611248889546891945355636218793299748845867834814 48626
980470399009162895288368911374616731561702410045157757001603778370339572539261354053250114657419224801218243169853035363945988575041132622240054109705194916292574379281916876204926756847774860345138396 94679
909435305586150042794448311206309059344324700093753671005604492655603472747780790783639504518762448392697987313535284348463888133230439792774802201529619235548600047342730950786780625307673643332282419 61324
270565577945226606569501618926256269482380056950336491504711624285796896829089690583637582782838928955203632239309604204622878106376581117827427625323405055645853432873936828810555823020155443402566605 61199

However, William Jones's work was far from influential in mathematical circles. In fact, the book probably would have been forgotten had he not made use of the symbol π. Another thirty years would pass before a major mathematician, Leonhard Euler, began using this single symbol in his work. In 1734, Euler was still using the letter *p* to specify π, and—for some reason—g to specify π/2. However, two years later, he began using π to denote the ratio of circumference to diameter in his papers and in correspondence.

No one knows whether Euler knew of Jones's use of π or not, but it hardly matters. As little as Jones's influence was, Euler's was great, and his use of the symbol π was contagious. In 1739, in a letter to Euler, Johann Bernoulli used the letter *c*, but by the very next year, he was using π. Nikolaus Bernoulli began using π in his correspondence with Euler soon after. And then, when Euler used the π notation in his *Introductio in analysin infinitorum* (published in 1748), the use of the symbol became widespread.

That's not to say that the circle's ratio was the only thing that the symbol π was used for during the eighteenth century. Pierre Hérigone used π to signify the proportion or ratio between two numbers. (Where we would write 2:3, he wrote 2π3.) A. G. Kästner used 1:*P* to denote the ratio of diameter to circumference, and used π to denote the circumference itself. His usage later became even more muddled when he used π in "cos *u* = π and sin *u* = *p*," and then π was the coefficient of the $(n+1)^{th}$ term of an equation. But by 1771, even Kästner had settled on the modern-day definition of π.

In 1799, Paolo Ruffini used the symbol π to signify factorials (he writes "π = 1.2.3.4 . . . . *m*"). In the years to come, however, this usage was to mutate slightly into the capital letter pi, Π. By 1811, Gauss, Jacobi, and Weber were all writing Π(*n*) for *n*-factorial. Then, Gauss also introduced the symbol Π to signify a continued product, which is still in use today.

Ultimately, by the time A. M. Legendre published his French textbook Éléments de géométrie in 1794, almost all mathematicians in Europe were using the symbol π in the same way that we do today.

π is currently also used to signify something besides the circle's ratio. Many mathematicians use π(n) as a function of the number of primes less than or equal to the number n. For instance, π(10) equals four, because there are four prime numbers between 1 and 10. Gauss, at age fifteen, discovered the famous prime number theorem:

$$\lim_{n\to\infty} \frac{\pi(n)}{(n/\log n)} = 1$$

Martin Gardner pointed out in his Wheels, Life, and Other Mathematical Amusements that the ancient yin-yang symbol can fool us into thinking that pi equals 2. If a circle has a radius of 1, then half the circle has a length of π. The lengths of the next two smaller semicircles total π, as do the next four smaller semicircles, and so on ad infinitum. However, at some point the semicircles get so small and so numerous that they approach the diameter of the original circle, which was 2. Therefore, pi must equal 2. Right? Wrong. Gardner explains: "At no step, however, do the semicircles alter their shape. Since they always remain semicircles, no matter how small, their total length always remains pi. The fallacy is an excellent example of the fact that the elements of a converging infinite series may retain properties quite distinct from those of the limit itself."

Q: What do you get if you divide the circumference of a jack-o-lantern by its diameter?

A: Pumpkin π

—John Evans

Most Macintosh computer programmers save their original source code with file names ending with the letter π (such as "mycode.π").

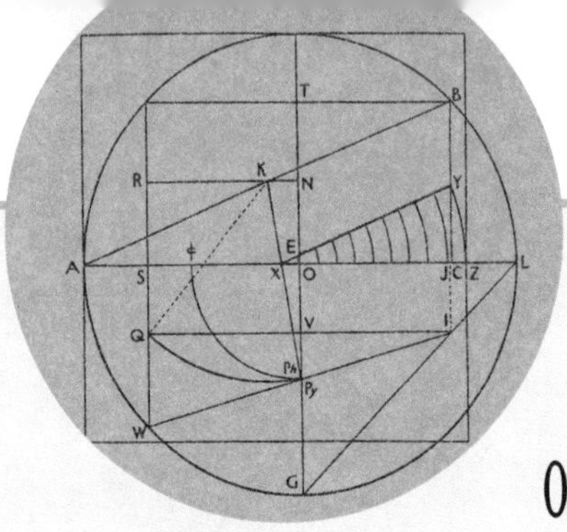

The People of the State of California, Plaintiff, v. Orenthal James Simpson, Defendant

[The following is a transcript of an interchange between defense attorney Robert Blasier and FBI Special Agent Roger Martz on July 26, 1995, in the courtroom of the O. J. Simpson trial.]

Q Can you calculate the area of a circle with a five-millimeter diameter?

A I mean I could. I don't . . . math I don't . . . I don't know right now what it is.

Q Well, what is the formula for the area of a circle?

A Pi R Squared.

Q What is pi?

A Boy, you are really testing me. 2.12 . . . 2.17 . . .

Judge Ito: How about 3.1214?

Q Isn't pi kind of essential to being a scientist knowing what it is?

A I haven't used pi since I guess I was in high school.

Q Let's try 3.12.

A Is that what it is? There is an easier way to do . . .

Q Let's try 3.14. And what is the radius?

A It would be half the diameter: 2.5.

Q 2.5 squared, right?

A Right.

Q Your honor, may we borrow a calculator?

[brief pause]

Q Can you use a calculator?

A Yes, I think.

Q Tell me what pi times 2.5 squared is.

A 19.

Q Do you want to write down 19? Square millimeters, right? The area. What is one tenth of that?

A 1.9.

Q You miscalculated by a factor of two, the size, the minimum size of a swatch you needed to detect EDTA, didn't you?

A I don't know that I did or not. I calculated a little differently. I didn't use this.

Q Well, does the area change by the different method of calculation?

A Well, this is all estimations based on my eyeball. I didn't use any scientific math to determine it.

The Personality of Pi

What good is your beautiful investigation regarding π? Why study such problems, since irrational numbers do not exist?

—**Leopold Kronecker to Ferdinand von Lindemann in 1882, the same year Lindemann proved the transcendence of pi**

DRABBLE® by Kevin Fagan

Numbers, like people, have qualities and characteristics—you can get to know a number and explore how it interacts with other numbers. Given a particular number, a mathematician might ask, Is it square? Is it prime? Can you transform the number into a continued fraction? Is the number short and sweet, ending with a dull thud after one or two decimal places, or does it roll infinitely along, spilling digits for an eternity? And if the latter is the case, do those digits endlessly repeat themselves in some sort of pattern, or does each number follow the next in what appears to be a random sequence?

We might also ask of an infinitely long number, "Is there a way to express you as a fraction of two integers?" Numbers that can be expressed in this way—positive or negative integers, fractions of those integers, and 0 (zero)—are called rational because you can easily express them as a ratio. Rational numbers, when converted into decimal notation, always end up repeating themselves sooner or later. If they're short—like 25/8, which equals 3.125—they terminate with an infinite number of zeros, signifying an exact measure. If they're infinitely long, the digits repeat in a cycle—as with 1/7, which equals .142857142857l4 . . . , a cycle of six digits.

So, what is the nature of the number π?

While mathematicians had believed for centuries that the number π was irrational (unable to be expressed as a ratio), it wasn't until 1761 that Johann Heinrich Lambert showed this conclusively. His method, while complex, boils down to the following argument: He first demonstrated that if x is a rational number, tan(x) must be irrational. Then, it follows that if tan(x) is rational, x must be irrational. Because tan(π/4) = 1, π/4 (and therefore π) must be an irrational number.

Some people felt Lambert's proof was not rigorous enough, but A. M. Legendre found another proof in 1794 that satisfied them; what's more, he proved that π^2 was irrational as well.

Once you've found a rational number, the next question you can ask of it is very telling about its character indeed: "Can you be expressed in an algebraic equation?" An algebraic equation always takes the form $a_n x^n + \ldots + a_2 x^2 + a_1 x + a_0 = 0$, where n is a finite number, and all coefficients (a_1, a_2, and so on) are rational numbers. Note

π
is the symbol Sandra Bullock clicks on to gain access to unauthorized databases on the Web in the movie The Net.

It is probable that the number π is not even contained among the algebraical irrationalities, i.e., that it cannot be a root of an algebraical equation with a finite number of terms, whose coefficients are rational. But it seems to be very difficult to prove this strictly. —A. M. Legendre, Éléments de géométrie, 1794

that a number can be irrational and still be algebraic. For instance, $x=\sqrt{2}$, which is irrational, can be easily expressed as $x^2-2=0$. Legendre felt that π could not be so reduced, but he could not prove his contention and died not knowing for sure. In fact, not until 1840, seven years after Legendre's death, did Joseph Liouville actually prove that numbers of this type—called transcendental numbers—actually exist.

Then, in 1873, Charles Hermite rigorously proved that the number e is transcendental. This only stoked the mathematical fires, heating up the discussion of whether π was truly transcendental as well. Hermite was once asked to continue his search in this direction. He politely declined, replying, "I shall risk nothing on an attempt to prove the transcendence of π. If others undertake this enterprise, no one will be happier than I in their success. But believe me, it will not fail to cost them some effort."

Nine years later, in 1882, Ferdinand von Lindemann proved that π is a transcendental number. His proof was built on foundations laid by two hundred years' worth of important mathematical contributions. Specifically, Hermite had shown that the number e was transcendental; that is, there is no equation $ae^m+be^n+ce^p+\ldots=0$, where the coefficients a, b, c . . . and the exponents m, n, p . . . are rational numbers. Lindemann then proved the more general theorem that a, b, c . . . , and m, n, p . . . also could not be algebraic numbers, not even nonreal numbers. This means that $e^{ix}+1=0$ cannot be satisfied when x is an algebraic number (we already know that i is algebraic). However, Euler had already established the equation $e^{i\pi}+1=0$. Therefore, π cannot be algebraic, and so it is transcendental.

One definition of pi: Twice some number between 0 and 2 whose cosine is 0.

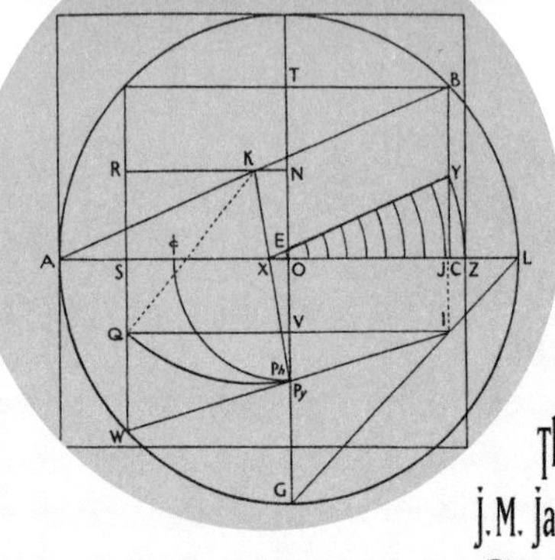

Pi Dialog

The following conversation is from Are Quanta Real? by J.M. Jauch (as quoted in Douglas Hofstadter's *Gödel, Escher, Bach*).

Salviati

Suppose I give you two sequences of numbers, such as 78539816339744830961566084 . . . and 1, -1/3, +1/5, -1/7, +1/9, -1/11, +1/13, -1/15, . . . If I asked you, Simplicio, what the next number of the first sequence is, what would you say?

Simplicio

I could not tell you. I think it is a random sequence and that there is no law in it.

Salviati

And for the second sequence?

Simplicio

That would be easy. It must be +1/17.

Salviati

Right. But what would you say if I told you that the first sequence is also constructed by a law and this law is in fact identical with the one you have just discovered for the second sequence?

Simplicio

That does not seem probable to me.

Salviati

But it is indeed so, since the first sequence is simply the beginning of the decimal fraction [expansion] of the sum of the second. Its value is $\pi/4$.

Simplicio

You are full of such mathematical tricks, but I do not see what this has to do with abstraction and reality.

Salviati

The relationship with abstraction is easy to see. The first sequence looks random unless one has developed through a process of abstraction a kind of filter which sees a simple structure behind the apparent randomness.

It is exactly in this manner that laws of nature are discovered. Nature presents us with a host of phenomena which appear mostly as chaotic randomness until we select some significant events, and abstract from their particular, irrelevant circumstances so that they become idealized. Only then can they exhibit their true structure in full splendor.

The Circle Squarers

One of the unnoticed good effects of television is that people now watch it instead of producing pamphlets squaring the circle.

—**Underwood Dudley,**
Mathematical Cranks

History is filled with attempts at changing one element of nature into another, whether through alchemy, magic, or science. Einstein, of course, provided some inspiration with his famous equation demonstrating how matter can be converted to energy. But for the most part, attempts at transmogrification were futile efforts at turning lead into gold—or, as the saying goes, making a silk purse from a sow's ear. However, along with these undertakings were some of a more geometrical bent, as people tried to turn circles into squares using every method they could dream up.

Squaring the circle is one of the few mathematical puzzles that has become commonly known outside of professional mathematical circles. Even if people on the street don't always know exactly what it means, they have probably heard of the problem and know that it is difficult or perhaps impossible. In fact, the phrase squaring the circle has filtered into our everyday language, suggesting a project that is doomed to failure.

To square the circle is to construct, either geometrically or numerically, a square with exactly the same area as a circle. (This is also called the quadrature of the circle.) The ancient Greeks, however, set up the circle-squaring problem with two conditions. First, the solution should use only a straightedge and compass (in order for the proof to be reduced entirely to Euclid's theorems). Second, the solution must be done without using an infinite number of steps. It turns out that it's easy to square the circle if you remove either of these constraints. For example, if you use higher mathematics, like calculus, or higher-order curves, like a quadratrix or a spiral, you can, in fact, construct a square of equal area to a circle quite easily.

Around 420 B.C.E., Hippias of Elis discovered a curve called a quadratrix, but it was not until 335 B.C.E. that Dinostratos used the quadratrix to build a square that had the same area as a circle. This didn't count as truly "squaring the circle" in the Euclidean sense, however, because the quadratrix requires an infinite number of steps to create.

Archimedes proved that the area of a circle equals that of a right triangle with one side equal to the circle's radius, and the second side equal to the circumference. Because of this, many people have attempted to square the circle by finding the circle's circumference based on its diameter (often called the rectification of the circle). Of course, if you know both the circle's diameter and circumference precisely, you know pi.

r

$2\pi r$

Augustus De Morgan dubbed St. Vitus the patron saint of circle squarers. In his *Budget of Paradoxes*, de Morgan writes that St. Vitus "leads his votaries a never-ending and unmeaning dance." He also points out that the saint is often represented alongside a rooster and adds, "Next, after *gallus gallinaceus* [the barnyard rooster] himself, there is no crower like the circle-squarer."

Nonetheless, two thousand years ago, no one knew that it would be impossible to pinpoint the ratio between circumference and diameter, and it became relatively common in ancient Greece to attempt squaring the circle. The Greeks even had a word, τετραγωνίζειν or *tetragonidzein*, that meant someone who busies himself with the quadrature of the circle.

But by the sixteenth century—not long after Cardinal Nicolas of Cusa declared that he had positively squared the circle (and was subsequently shown the error of his ways)—mathematicians were becoming aware that attempting to square the circle was futile. Here, perhaps, is where the split between true students of mathematics and the more common "circle squarers" began. Where Viète, Snell, Wallis, Newton, and others took the high road of incrementally understanding the infinite nature of pi, the circle squarers stubbornly held to the seductive belief that if they just worked hard enough, they would be able to solve—once and for all—this age-old problem.

Orontius Fineus, the great philologist Joseph Scaliger, Longomontanus of Copenhagen, and Gregory of St. Vincent are among the many mathematicians of the sixteenth and seventeenth centuries whose erroneous "proofs" were successfully refuted one by one. Even the well-known English philosopher Thomas Hobbes demonstrated his quadrature of the circle in his *De Problematis Physicis* (1668), where he shows that pi equals 3 1/8 (a common solution by circle squarers). Later, after his "proof" was clearly rejected, Hobbes attempted to defend his ludicrous position by disputing the fundamental principles of geometry as well as the Pythagorean theorem.

By 1775, so many people were trying to get validation for their method of squaring

The end of circle-squaring will come only as a consequence of the end of civilization.
—Underwood Dudley, *Mathematical Cranks*

the circle that the French Academy of Sciences passed a resolution not to examine any further so-called solutions of the quadrature of the circle. At the same time, the academy also excluded several other impossible tasks: the duplication of the cube, the trisection of the angle, and perpetual motion machines (devices that create as much energy as they take to run, enabling them to run perpetually).

In fact, trying to square the circle is very similar to attempting to find a method of perpetual motion. At first glance, each appears to be a problem that *should* have a solution, if only one could think it through well enough. It's this sort of reasoning that has been the impetus for people around the world to suddenly become circle squarers. Many of these people labor under the mistaken belief that squaring the circle is a central problem in mathematics. Some have even gone so far as to dub it the great end and object of geometry, and contend that if this one problem were solved it would be a great leap forward for humankind. Perhaps it would be, if only it were possible (which, of course, it's not).

Another misconception is that there is a grand monetary prize awaiting anyone who can successfully square the circle. In his book *A Budget of Paradoxes*, nineteenth-century scientific critic Augustus De Morgan wrote of a Jesuit who traveled from South America to England in order to claim his reward for squaring the circle, and of a certain M. de Vausenville, who brought legal action against the French Academy of Sciences in order to recover a reward which he

The famous English poet Alexander Pope wrote, in his 1743 poem *Dunciad*:

Mad Mathesis alone was unconfined,
Too mad for mere material chains to bind,—
Now to pure space lifts her ecstatic stare,
Now, running round the circle, finds it square.

Pope's author's notes explain that this refers to "the wild and fruitless attempts of squaring the circle."

deemed was his by way of his "solution." Of course, even if the methods had been true (which they could not have been), there had never been a reward offered for the solution in the first place. Even today, there are frequent reports from around the globe that someone somewhere has found a solution to the circle-squaring problem.

However, most cyclometers (those who are taken by the measure of the circle) are hell-bent on the belief—incredible as it may seem—that they have found the key to an ancient puzzle. And then, with the passion of those who have suddenly found religion, they proselytize and defend their "truth" to anyone who will listen.

Underwood Dudley, in his delightful book Mathematical Cranks, writes about a certain J. V., who, in November 1982, wrote that he had calculated pi to be "exactly 3.0625." Then, a month later, he wrote:

> Being the ONE "Quadrature of a Circle" the highest mathematical achievement of mankind, ever, and I, myself, with a profound sense of its scientific importance, for one and for all, in a definite final way do attest the "Quadrature of a Circle" as a true FACT.
>
> The first man ever to work the area of a circle, mathematically, without the value of "PI," I do further attest that the true one value of the proportion of the circumference of a circle to its diameter is "PI," that is, the total value of 2.91421351481511+, no buts, no more than a little less, period.

Clearly, one of the most fascinating traits of the circle squarer is a resilient nature that can withstand even the most profound arguments refuting his or her stance. At times, the cyclometer holds the faith that his solution is correct like a sword against reason. For example, in his book *The Quadrature of the Circle*, published in 1874, John A. Parker rationalized that "the circumference of a circle is the line outside of the circle thoroughly enclosing it," and then pointed out that everyone else's values for pi are wrong because "by this difference, with their approximation, geometers make an error in the sixth decimal place."

Some of these rationalizations can be convincing, although they remain unsubstantiated. Many cyclometers who carefully draft geometric solutions (using only compass and straightedge, without formulas) argue that their methods are more pure than numbers because they are working on a higher plane of reasoning. Therefore, when you use their solution to try something as simple as finding the circumference of a circle with a diameter of 1, they become sullen and insist that working with numbers just muddies the water. Rather it appears that the muddy water is found not in the digits but between their ears.

Sometimes, these geometric cyclometers insist—with a straight face—that the numerical definition of pi may be 3.14159265 . . . , but that their geometric answer is *also* correct.

Those circle squarers who attempt to calculate pi themselves may find—lo and behold—a value for pi different from the one that is widely accepted. Many of these people check and recheck their work (usually repeatedly making the same mistake they made in the first place), and finally make the proclamation that they hold the correct value and every great mathematician and scientist today is wrong.

One of the numerical circle squarers' favorite techniques is the *reductio ad absurdum*, in which you make an assumption at the beginning of the proof and then show why any other answer would be absurd. For instance, they assume that pi equals 3.125, and then go on to show that the accepted value of 3.1415+ would be impossible. Unfortunately, the circle squarers' faulty proofs are too often constructed in such a way that the original value is correct no matter what value they insert. So even if they had first made the assumption that pi equaled 47, they would have "discovered" that pi was much larger than had been previously thought—more than fifteen times larger, not surprisingly.

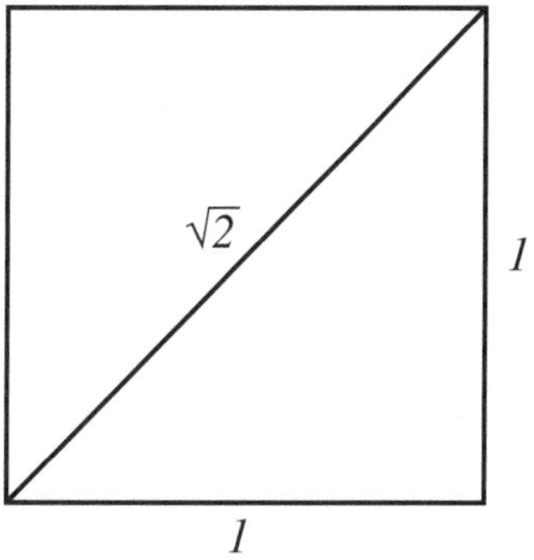

True, the proof that pi is irrational (it can't be expressed as a ratio of integers, as Lambert proved in 1761) was a blow to some circle squarers, but it did not prove fatal because you can still construct some irrational numbers geometrically. For instance, it's relatively easy to draw, using only straightedge and compass, the relationship between the side and the diagonal of a square, even though this is an irrational number ($\sqrt{2}$).

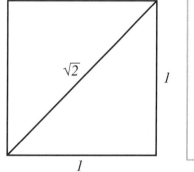

In fact, it wasn't until 1882—when Lindemann proved that pi is a transcendental number—that circle squaring could conclusively be put to rest. It had already been shown that only algebraic formulas that can be reduced to the second order can be built geometrically with straightedge and compass. When Lindemann proved

The Squaring of the Circle is of great importance to the geometer-cosmologist because for him the circle represents pure, unmanifest spirit-space, while the square represents the manifest and comprehensible world. When a near-equality is drawn between the circle and square, the infinite is able to express its dimensions or qualities through the finite. —Robert Lawlor, Sacred Geometry, 1982

pi's transcendental nature, he showed that there was no finite algebraic equation that could describe the number, and therefore it could not possibly be constructed within the classical Euclidean constraints.

But once again, attempting to convince someone whose mind is already made up is difficult. In Mathematical Cranks, Underwood Dudley also tells of a cyclometer who wrote that "π's only position in mathematics is its relation to infinite series [and] that π has no relation to the circle. . . . Lindemann proclaimed the squaring of the circle impossible; but Lindemann's proof is misleading for he uses numbers (which are approximate in themselves) in his proof." How can you argue with that logic?

And then there is the type of circle squarer for whom a good ruler and a careful measurement are enough to satisfy a thirst for absolute truth. For instance, Augustus De Morgan tells the story of the man who declared, "I thought it very strange that so many great scholars in all ages should have failed in finding the true ratio, and have been determined to try myself." The man went on to roll a disk 12 inches in diameter along a ruler, finding that the ratio of circumference to diameter is "exactly" 3.140625.

The problem with arguing with this sort of circle squarer is that he can show you that he is right. True enough, it is difficult to match the precision even of 22/7 (3.143) in most real-world mechanical work that involves circles, and if you had to do the math longhand, you'd find that you'd try to get by with 3 1/8 (3.125), which is significantly easier to calculate.

Of course, circle squarers maintain an arsenal of excuses as to why prominent mathematicians so quickly dismiss their solutions. One conspiracy theory argues that mathematicians who have written textbooks are bound to lose money if they're proven wrong; therefore, they'll fight any argument. Other cyclometers have more than a healthy dose of paranoia and believe that there are mathematicians and scientists who have been keeping the "real value of pi" from us—perhaps because, like Freemasons, they feel the need to keep the truth for themselves.

And many other cyclometers simply think that the mathematical community just isn't fit for such important undertakings. As circle squarer Lawrence Cavender wrote in his 1967 work Unique Mathematical Geometrical Findings, "Why did not the mathematicians discover these truths in the past? Primarily; because they did not approach these solutions in the proper manner. Secondly; no one dared to even consider that it was possible for great mathematicians to have erred in these matters." Cavender, of course, had dared and had approached the problem, but to little avail in the long run.

It must be extremely difficult for circle squarers to argue with the establishment day after day, insisting that they are correct and that all other mathematicians are patently wrong. Of course, there's a certain pride involved in standing up as an individual against an institution, hoping to demonstrate that anyone with a good idea and perseverence can rise to the top. Unfortunately, when it comes to mathematics, the circle squarers' "good idea," mixed with their industrial-strength perseverence, tends to be a better recipe for ridicule than for recognition.

When we think of π, let's not always think of circles. It is related to all the odd whole numbers. It also is connected to all the whole numbers that are not divisible by the square of a prime. And it is part of an important formula in statistics. These are just a few of the many places where it appears, as if by magic. It is through such astonishing connections that mathematics reveals its unique and beguiling charm. —Sherman K. Stein, Strength in Numbers

[The circle squarer] is determined to secure recognition, and appeals therefore to the public. The newspapers must obtain for him the appreciation that scientific societies have denied. And every year the old mathematical sea-serpent more than once disports itself in the columns of our newspapers in the shape of an announcement that Mr. N.N., of P.P., has at last solved the problem of the quadrature of the circle. —Hermann Schubert, Mathematical Essays and Recreations (translated by Thomas McCormack)

The Sophists of ancient Greece were taken with the idea of squaring the circle. However, they insisted that the answer depended on finding a number that represented both a square and a circle. A square, meaning a square of a second number; and a circle, meaning that the number would end in the same digit as that second number. For instance, 36 is a square of 6 and ends in the number 6.

Squaring the Circle is the title of two plays: one by the contemporary playwright Tom Stoppard, first produced in 1984, and the other by the Russian dramatist Valentin Katayev. Originally produced at the Moscow Art Theatre in 1928, it was Katayev's most successful play.

The race of circle-squarers, unmindful of the verdict of mathematics, the most infallible of arbiters, will never die out as long as ignorance and the thirst for glory remain united. —Hermann Schubert, Mathematical Essays and Recreations, 1898 (translated by Thomas McCormack)

Meton: With the straight ruler I set to work
To make the circle four-cornered
—Aristophanes, The Birds (414 B.C.E.)

Let us all give thanks, loudly and daily, that we are not cranks devoting all of our energies to convincing the world that π is exactly 3.125, 22/7, or $\sqrt{10}$. The chance that our lives will be rich, delightful, and full of meaning may be small, but it is greater than the near-zero chance of the circle-squarer. —Underwood Dudley, Mathematical Cranks

As the geometer who sets himself to measure the circle and who findeth not, think as he may, the principle he lacks, such was I at this new seen spectacle. —Dante Alighieri, Paradiso, canto 33

It should seem that it is easier to square the circle than to get around a mathematician. —Augustus De Morgan, A Budget of Paradoxes

One would almost fancy that amongst circle-squarers there prevails an idea that some kind of ban or magic prohibition has been laid upon this problem; that like the hidden treasures of the pirates of old, it is protected from the attacks of ordinary mortals by some spirit or demoniac influence, which paralyses the mind of the would-be solver and frustrates his efforts. —John Phin, The Seven Follies of Science, 1912

On the eighteenth-century claim that squaring the circle is key to understanding the longitude problem, Augustus De Morgan wrote in his Budget of Paradoxes: "Sometimes a cyclometer persuades a skipper, who has made land in the wrong place, that the astronomers are in fault for using a wrong measure of the circle; and the skipper thinks it a very comfortable solution! And this is the utmost that the problem ever has to do with longitude."

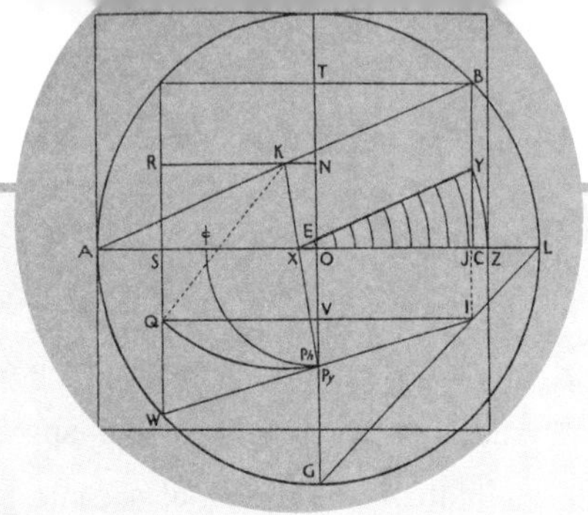

From the Mouths of Circle Squarers

Whenever I have forced opponents into a difficulty, they have almost invariably attempted to escape it by pleading want of time to enter into a correspondence; the day, however, is not far distant when it will be universally admitted, that my conclusion, vis., "25/24 (circumference) in every circle, is exactly equal to the perimeter of an inscribed regular hexagon," never has been refuted, and never can be; and it, therefore, follows, <u>of necessity</u>, that 1 to 3 1/8 is the true and exact ratio of diameter to circumference in every circle.

—James Smith, <u>The Quadrature and Geometry of the Circle Demonstrated</u>, 1872

I have found, by the operation of figures, that this proportion is as 6 to 19. Now, in order to make a ratio, I divide the 19 by 6, which gives 3.166 4/6 . . . . I am asked what evidence I have to prove that the proportion the diameter of a circle has to its circumference is as 6 to 19? I answer, there is no other way to prove that an apple is sour, and why it is so, than by common consent.

—John Davis, <u>The Measure of the Circle</u>, 1854

The circumference of any circle being given, if that circumference be brought into the form of a square, the area of that square is equal to the area of another circle, the circumscribed square of which is equal in area to the area of the circle whose circumference is first given.

—John A. Parker, The Quadrature of the Circle, 1874

This measure will and must prove a great benefit to mankind, when understood, as it is the basis and foundation of mathematical operations; for, without a perfect quadrature of the circle, measures, weights, etc., must still remain hidden and unrevealed facts, which are and will be of great importance to rising generations. The improvements that will arise from this measure fifty years hence I cannot paint in imagination.

—John Davis, The Measure of the Circle, 1854

The British Association for the Advancement of Science may assume infallibility, and authoritatively proclaim that the solution of the problem is impossible; and may consequently decline to permit the consideration of the subject to be introduced into their deliberations. . . . And yet, the solution of the problem is extremely simple after all. It would almost appear as if its very simplicity had been the grand obstacle which had hitherto stood in the way of its discovery. . . . I have subjected my theory to every conceivable test, both mathematical and mechanical, with an honest determination to find a flaw if possible; and having failed to do so, I now unhesitatingly propound it, as the true theory on this important question.

—James Smith, The Quadrature of the Circle, 1861

[The proof of the irrationality of pi] definitely disposed of the problem of squaring the circle, without, of course, dampening in the least the ardor of the circle-squarers. For it is characteristic of these people that their ignorance equals their capacity for self-deception. —Tobias Dantzig, Number: The Language of Science, 1930

It is utterly impossible for one to accomplish the work in a physical way; it must be done metaphysically and geometrically, not mathematically. When approached in this manner the problem is easy of solution. It is stated by occultists that the number 12 squares the circle and it is necessary to take into consideration the process before we can understand this; when correctly understood we know it to be perfectly true.

—A. S. Raleigh, Occult Geometry, 1932

The original conception, its natural harmony, and the result, to my own mind is a demonstrative truth, which I presume it right to make known, though perhaps at the hazard of unpleasant if not uncourteous remarks.

—James Sabben, after declaring that pi equals 3.14176 in his Method to Trisect a Series of Angles Having Relation to Each Other, 1848

Now, being in my 81st year . . . I found time to set my thinks to work. After three long days I was divinely blessed to discover the true and ancient value of Pi. This is the exact value used by the Great Creator when he created our earth and its four orbit cycles. The World's unsolved problem, i.e., the complete interlocking of a triangle, circle, and square—having equal areas—is in the construction of this Pyramid. This truly confirms that the Great Creator used the Pi formula of 3 (7.1)/(50)=3.1420000. The very fact that the Great Creator, Archangel Michael, used this 3.1420000 value of Pi in His creating our earth and its orbits, should be sufficient evidence for our adoption.

—Francis Michael Darter, The Story of True Pi, 1962

No amount of attestation of innumerable and honest witnesses, would ever convince anyone versed in mathematical and mechanical science, that a person had squared the circle or discovered perpetual motion. —Baden Powell, Essays and Reviews, 1860

Augustus De Morgan tells us in his Budget of Paradoxes that the eighteenth-century Frenchman Joseph-Louis Vincens de Mouléon de Causans "cut a circular piece of turf, squared it, and deduced original sin and the Trinity. He found out that the circle was equal to the square in which it is inscribed, and he offered a reward for detection of any error, and actually deposited 10,000 francs as earnest of 300,000. But the courts would not allow anyone to recover."

Surely the cyclometer is a Darwinite development of a spider, who is always at circles, and always begins again when his web is brushed away. —Augustus De Morgan, A Budget of Paradoxes

One nineteenth-century circle squarer, after praising God repeatedly for choosing him to reveal "this precious mathematical jewel," goes on to show that an inscribed square can be broken down into four congruent triangles, and a circumscribed square is made of eight. Therefore, he asserts, the area of the circle must be the area of six triangles.

Nineteenth-century writer James Smith surely holds the record for most publications on the topic of circle squaring. However, it should be noted that he paid for each and every one to be printed. Curiously, he always argued the same point: that pi is $3\,^1/_8$. Augustus De Morgan once wrote of him, "He is beyond a doubt the ablest head at unreasoning, and the greatest hand at writing it, of all who have tried in our day to attach their names to an error. Common cyclometers sink into puny orthodoxy by his side. . . . We can only say this: he is not mad. Madmen reason rightly upon wrong premises: Mr. Smith reasons wrongly upon no premises at all."

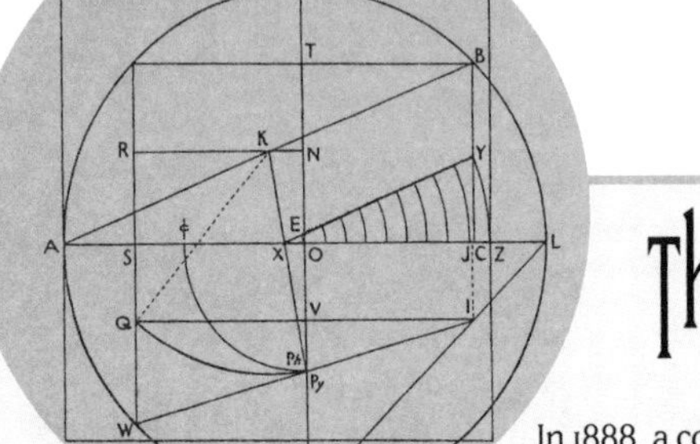

The Legal Value of Pi

In 1888, a country doctor named Edwin J. Goodwin, who lived in a small town named Solitude, Indiana, "discovered" a solution to the circle-squaring problem. Actually, he claimed that he had been "supernaturally taught the exact measure of the circle . . . in due confirmation of scriptural promises." But it wasn't until nine years later, in 1897, that he was able to persuade his state representative, Taylor Record, to introduce a bill into the Indiana state legislature. This bill offered the state—on condition of the bill being passed—the right to have free use of Goodwin's mathematical findings, whereas other states would presumably be required to pay a royalty for the work. (Of course, you cannot copyright a mathematical truth anyway, so this offer was meaningless.)

House Bill No. 246—probably written by Goodwin himself, for Record later claimed to have no idea what the bill contained—would not have actually required the Indiana educational system to use Goodwin's value for pi, though the implication and perception would certainly have been that Indiana both acknowledged and endorsed Goodwin's findings. And although the bill ultimately failed, it's fascinating that it progressed as far as it did in the state congress.

When the bill first entered the Indiana house of representatives, it was quickly and mysteriously referred to the House Committee on Canals, generally referred to as the Committee on Swamp Lands. The bill did not stay long in the swamp, however, and it was passed to the Committee on Education, which turned it around the same week with a recommendation that it pass.

Clearly, none of the lawmakers understood the mathematical contents of the bill or knew that what Goodwin was proposing was impossible. What with the busy schedules of the representatives, this isn't surprising. What <u>is</u> odd is that the state superintendent of public instruction was one of the backers of the bill as well.

Also strange was that the bill included a reassuring comment that Goodwin's calculations were correct: "In further proof of the value of the author's proposed contribution . . . is the fact of his solutions of . . . the quadrature of the circle having been already accepted as contributions to science by the American Mathematical Monthly, the leading exponent of mathematical thought in the country." True, Goodwin's work had been printed in the journal. Unfortunately, the bill did not state that alongside the author's article was printed "Published by the request of the author." Perhaps the journal had some additional space to fill, or perhaps Goodwin simply badgered the editors into printing it. Whatever the case, it is certain that the journal did not intend an endorsement.

Without knowing the whole truth, the bill was soon passed unanimously by the house of representatives (67 to 0). The next day, the Indianapolis Journal, reporting on the event, wrote that this was the strangest bill that had ever passed an Indiana assembly.

Apparently, C. A. Waldo, a professor of mathematics at Purdue University, was visiting the house of representatives the day the bill was passed and stumbled upon the discussions. If he was surprised by a discussion on mathematics in this place of law, he was even more surprised when he found that Dr. Goodwin and his findings were being lauded. He later wrote that he declined an offer to meet the "learned doctor" that day, noting that "he was acquainted with as many crazy people as he cared to know."

Fortunately, when the bill was passed to the state senate floor for confirmation, it became caught in some bureaucratic bungling and was passed to the Committee on Temperance. By the time the committee's chairman made a recommendation to the senate that the bill be passed, the national news media had begun to run articles about it. Between the counseling of Dr. Waldo and the increasingly sarcastic tone of the local newspaper editorials, the senators quickly decided to dispose of the bill.

Ultimately, their public explanation had nothing to do with the merits of the bill. Rather, they curtly noted that this was not a proper subject for legislation, and on February 12, 1897, the senators voted to postpone any further discussion of it indefinitely. To this day, the subject has not surfaced again, proving once again that it's easier for the government to say "maybe" than it is to say "no."

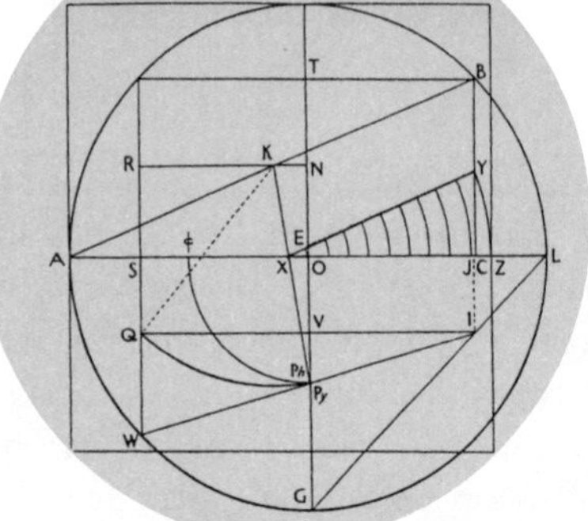

Picture a large rectangular grid of sticks, planted upright in the ground, like an orchard of trees. If you pick any two of these sticks at random, the probability that you would be able to see one from the other (without any other stick getting in the way) is $6/\pi^2$.

'Tis a favorite project of mine
A new value of pi to assign.
I would fix it at 3
For it's simpler, you see,
Than 3 point 1 4 1 5 9.

—Harvey L. Carter, quoted by
W. S. Baring-Gould, in
The Lure of the Limerick

These four numbers are prime (a prime number can only be divided without a remainder by itself and 1):3

31

314,159

31,415,926,535,897,932,384,626,433,832,795,028,841 (found to be prime by Robert Baillie and Marvin Wunderlich at the University of Illinois in 1979)

Curiously, if you reverse the order of the first three of these, you also get prime numbers:

3

13

951413

Of course, if we know the area of the circle, it is easy to find the side of a square of equal area; this can be done by simply extracting the square root of the area, provided the number is one of which it is possible to extract the square root. Thus, if we have a circle which contains 100 square feet, a square with sides of 10 feet would be exactly equal to it. But the ascertaining of the area of the circle is the very point where the difficulty comes in; the dimensions of circles are usually stated in the lengths of the diameters, and when this is the case, the problem resolves itself into another, which is: To find the area of a circle when the diameter is given. —John Phin, The Seven Follies of Science, 1912

Choose any two integers. The probability that they are relatively prime (they have no common factors) is $6/\pi^2$.

It was Augustus De Morgan who first coined the phrase morbus cyclometricus (the circle-squaring disease), in his Budget of Paradoxes.

Math Poem

$$\int_1^{\sqrt[3]{3}} z^2 dz \times \cos\left(\frac{3\pi}{9}\right) = \ln(\sqrt[3]{e})$$

Which, of course, translates to:

Integral z-squared dz
from 1 to the cube root of 3
times the cosine
of three pi over 9
equals log of the cube root of e.

—Anonymous

In his Mathematical Essays and Recreations (1898), Hermann Schubert wrote about the eighteenth-century Frenchman Oliver de Serres, "who by means of a pair of scales determined that a circle weighed as much as the square upon the side of the equilateral triangle inscribed in it, that therefore they must have the same area, an experiment in which $\pi = 3$."

Memorizing Pi

Hold this sign up to a mirror.

Pi is not just a collection of random digits. Pi is a journey; an experience; unless you try to see the natural poetry that exists in pi, you will find it very difficult to learn.

—Antranig Basman

There's a good reason most telephone numbers in the world are between six and eight numbers: We humans aren't very good at remembering large chunks of information. In fact, we typically have trouble recalling any more than seven or eight discrete pieces of information at a time. You might remember that there were ten cars parked in a lot, but if they were all different, you'd probably be hard-pressed to remember each one's color and its order in line. You might even remember your credit card number, but you'll probably recall it as four groups of four numbers each rather than sixteen digits in a row.

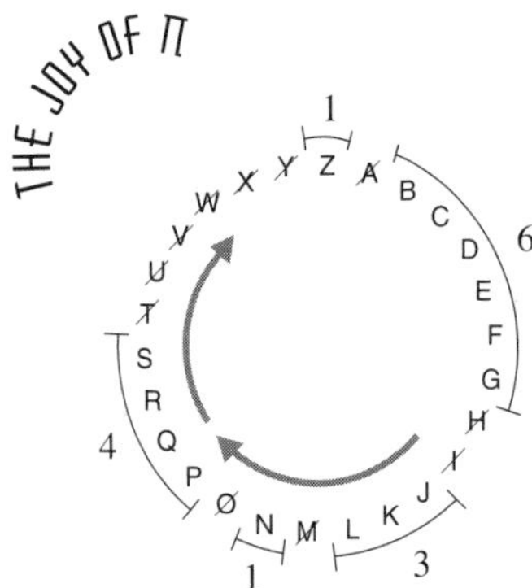

Martin Gardner once pointed out that when you cross out all the symmetrical letters in an alphabet circle, the resulting letters group into digits of pi.

How and what we remember is a mystery to the sciences, both neurological and psychological. Why can we hold ten individual telephone numbers in mind, but have great difficulty storing the same seventy digits all in a row? How is it that we can recall a long story, even one that contains dozens of events, but we can hardly remember ten words picked at random for more than a few seconds?

Many people simply accept their limited memory and are content to ask, "Where did I put my keys again?" Others find their memory a noble adversary and challenge themselves to memorize as much data as they can. People memorize decks of 52 shuffled playing cards; they memorize numbers from a page of a telephone book; they memorize sports scores and statistics. But perhaps most of all, people looking for a challenge try to remember numbers, and the single most popular number to memorize is pi.

High school students have long tried to memorize pi to 25, 50, or even 100 digits to impress their friends. That may seem like a lot, but remember that actors are asked to learn pages and pages of text; it's not so hard with a few days of practice. However, there are those who take these feats of memory to a whole different level.

Alexander Craig Aitken, professor at the University of Edinburgh and one of the greatest all-around mental calculators of our time, has an exceptional memory, and once memorized pi to 1,000 digits on a lark. Mathematician and author Martin Gardner writes in his Whys and Wherefores that Aitken "distrusted all mnemonic tricks. 'They merely perturb with alien and irrelevant association a faculty that should be pure and limpid' [Aitken said.] His way of memorizing pi was to arrange the digits in rows of fifty each, then divide each fifty into ten groups of five and 'read all these off in a particular rhythm. It would have been a reprehensibly useless feat, had it not been so easy.' "

Twenty years ago, Simon Plouffe held the world record for memorizing pi. Plouffe, who told people that he had memorized 4,096 digits, had actually memorized 300 more than that. But, being a mathematician, he thought that 4,096 (which is 2^{12}) sounded better. When asked why he had memorized so many digits, he simply replied, "I was young and I had not much else to do, so I did it." Ivars Peterson, another pi lover, later wrote, "To Plouffe, memorizing the digits of pi was close to a mystical experience. . . . To preserve the numbers in long-term memory, he periodically isolated himself in a room—no lights, no noise, no coffee, no cigarettes. 'Like a monk,' Plouffe says. As he recited the digits to himself, they would gradually seep into his mind."

Rajan Mahadevan is even more extraordinary when it comes to remembering numbers. He set the 1983 world record while in his twenties by reciting 31,811 digits of pi, but he is equally adept at memorizing license plate numbers, railway timetables, or any other list of information. Ironically, years later, as a graduate student at Kansas State University, his roommate revealed to a reporter from People magazine, "He's very good at remembering phone numbers, but I have to tell him over and over again how to work the VCR."

But none of these people hold a candle to twenty-three-year-old Hiroyuki Goto, who in February 1995 spent just over nine hours reciting 42,000 digits of pi from memory.

In April 1995, the Reuters wire service reported that a twelve-year-old Chinese boy, Zhang Zhuo, recited the value of pi to 4,000 decimal places from memory. Apparently, this took him just over twenty-five minutes.

For most of us, it would be hard work to memorize a thousand of anything, much less a list of 42,000 apparently random numerals. But there are techniques that any of us can use to memorize pi. The most common method is the word-length mnemonic, in which the number of letters in each word equals a digit of pi.

Perhaps the most simple of these mnemonics is "How I wish I could calculate pi." Of course, if you're the type to round numbers, you may want to replace the last word with pie, resulting in 3.141593. Similarly, college students for years have found 15 digits of pi using "How I like a drink, alcoholic of course, after the heavy lectures involving quantum mechanics." But to really stretch out the digits, you'll have to use a poem, such as this one:

Sir, I bear a rhyme excelling
In mystic force and magic spelling
Celestial sprites elucidate
All my own striving can't relate.
Or locate they who can cogitate
And so finally terminate.   Finis.

This device takes you out to pi's 31st decimal digit (3.1415926535897932384626433832795). At digit number 32, these mnemonics encounter a roadblock: the digit 0. We can only assume that a word with zero letters, when read out loud, sounds like one hand clapping. Nonetheless, various die-hard authors have developed methods for dealing with zero—usually using a ten-letter word, or placing some form of punctuation where the zero should appear.

Have a look at the digits 151–180. Imagine that you want to memorize them all.
481117450284102701938521105559
Which digits are the easiest? I think it's 111 and 555. Now we've split one big problem into five smaller ones:
48-111-745028410270193852110-555-9
It starts with 48, then comes the first easy number 111, then a medium problem, then the second easy number 555, and finally it ends with a 9.
Now analyze the "medium problem."
I split it the following way:
7450 28410 270 (three groups ending with a zero)
1938 (a year number)
52 (the number of weeks in a year)
110 (the emergency telephone number in Germany)
The main problem is to find patterns and divide the digits into the right groups. Then memorization can be done in very little time. Try it out!

—Mark Dettinger

For example, Mike Keith wrote a pi-mnemonic poem that not only encodes the first 740 digits of pi but is carefully crafted to reflect Edgar Allan Poe's "The Raven." The following represents only 80 digits of the total (the entire poem can be found at www.joyofpi.com). Note that words with fewer than ten letters represent the number of letters (e.g., pie = 3), ten-letter words equal the number 0, and words with more than ten letters equal two digits (e.g., a twelve-letter word represents 1 followed by 2). Ignore all punctuation.

> Learning many digits of pi carries with it certain severe dangers, and can do mental damage that takes a long time to heal. The mind knows perfectly well that learning digits of pi is not something to its benefit, and is likely to take severe retaliation against you once it works out what you are doing. . . . If you start to feel these effects, stop learning immediately.
>
> —Antranig Basman

Pie, E.: Near a Raven

Midnights so dreary, tired and weary.
Silently pondering volumes extolling all by-now obsolete lore.
During my rather long nap—the weirdest tap!
An ominous vibrating sound disturbing my chamber's antedoor.
"This," I whispered quietly, "I ignore."

Perfectly, the intellect remembers: the ghostly fires, a glittering ember.
Inflamed by lightning's outbursts, windows cast penumbras upon this floor.

Sorrowful, as one mistreated, unhappy thoughts I heeded:
That inimitable lesson in elegance—Lenore—
Is delighting, exciting . . . nevermore.

However, word-length mnemonics still only take you so far when memorizing pi. To memorize pi into the thousands of digits, Michael Harty used a phonetic code in which each consonant sound represented a number from 0 to 9. The sounds, when placed together, form words, and the words form long surreal stories.

For example, the sounds for the second and third digits—1 and 4—are t and r, which together sound like tear (as in teardrop). The next three—1, 5, and 9—are t, l, and p, which sound like tulip. Using this method, Harty memorized over 7,000 digits of pi and can repeat them by simply recalling a story (albeit an odd one): "[the 3 is] crying on a tulip, which moves away and picks up a device which puts a big notch into a yellow mule. The yellow mule becomes quite upset and he grabs a big box of laundry detergent called Fab . . . and he dumps it on a gray bomb and the gray bomb, of course, gets angry as well. . . ."

Some people have noted that learning pi in this way may be significantly easier in languages such as Japanese. When you read the numbers of pi out loud in Japanese—

If you are musically inclined, you might want to memorize the digits of pi as though they were musical notes. For instance, if 1 = C, 2 = D, 3 = E, and so on, then:

san ichi yon ichi go ku ni roku go san hachi—the combination of some numbers is the equivalent of other Japanese words: *ichigo* is *strawberry*, *kuni* is *country*, and so on. By creating a mental picture of the scene, it's easy to remember "3.14 strawberry country at 6 A.M., five chopsticks . . ." (Of course, this includes a modicum of fudging. For example, *chopsticks* is the translation of *hashi* rather than *hachi*, which means *eight*.)

> Pi is an ideal number for an exhibition such as [memorizing numbers] because it has no pattern to its digits.
>
> —Gregory Pinney, *Minneapolis-St. Paul Star Tribune*

And Mandarin Chinese may be even easier, for each number is a single syllable. When read out loud, the first twenty-eight digits actually sound like a poem! (Unfortunately, the poem doesn't say anything . . . but that's not uncommon in math poetry.)

san yi si yi wu jiou er liou
wu san wu ba jiou qi jiou
san er san ba si liou er
liou si san san ba san er

Ultimately, it's clear that memorizing digits of pi is little more than a great stunt for cocktail parties. But is that such a bad thing? When asked why he spent the time and effort to memorize pi, Alexander Volokh answered, "Obviously, no one needs to remember 167 digits of pi. Most people don't quite see the utility of it, but that's okay. To say that math has to be useful is like saying the English language is only good for ordering pizza."

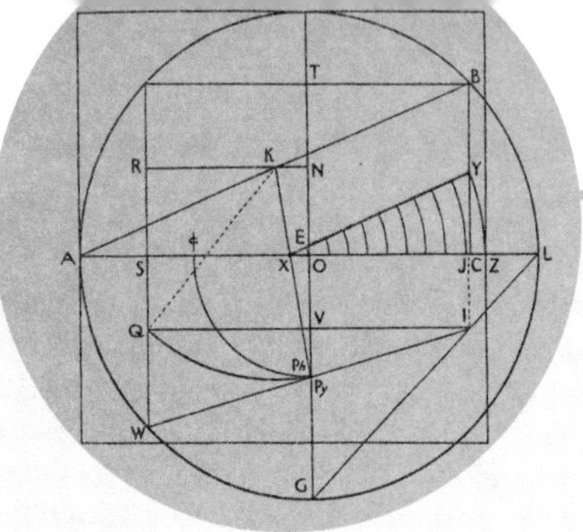

Pi Mnemonics from Around the Globe

Dutch

Eva o lief, o zoete hartedief uw blauwe oogen zyn wreed bedrogen.

(Eve, oh love, oh sweet darling your blue eyes are cruelly deceived.)

—Popular song at the Mathematics and Physics Department, University of Nijmegen

English

See, I have a rhyme assisting
my feeble brain,
its tasks oft-times resisting.

—Anonymous

If a billion digits of pi were printed in ordinary type, the expression would extend over 1,200 miles.

French

Que j'aime à faire apprendre ce nombre utile aux sages!
Immortel Archimède antique, ingénieur,
Qui de ton jugement peut sonder la valeur?
Pour moi ton problème eut de pareils avantages.

(How I like to teach this useful number to the wise!
Immortal Archimedes, artist, engineer.
Who, in your judgement, can grasp the value?
To me, your problem has similar advantages.)

—Anonymous

Greek

Αει ο Θεος ο Μεγας γεωμετρει
Το κυκλου μηκος ινα ορισῃ διαμετρω
Παρηγαγεν αριθμον απεραντον
και ον φευ ουδεποτε ολον
θνητοι θα ευρωσι.

(Great God ever geometrizes
To define the circle length by its diameter
Produced an endless number
Which whole, alas, mortals
Will never find.)

—Nikolaos Hatzidakis

ITALIAN

Che n' ebbe d' utile Archimede da ustori vetri sua somma scoperta?

(What good came to Archimedes from his immense discovery of burning mirrors?)

—ISIDORO FERRANTE

SPANISH

Sol y Luna y Mundo proclaman al Eterno Autor del Cosmo.

(Sun and Moon and World acclaim the eternal author of the Cosmos.)

—DAVID LANTZ

SWEDISH

Ack, o fasa, π numer förringas
ty skolan låter var adept itvingas
räknelära medelst räknedosa
och så ges tilltron till tabell en dyster kosa.
Nej, låt istället dem nu tokpoem bibringas!

(Oh no, π is nowadays belittled
for the school makes each student learn
arithmetic with the help of calculators
and thus the tables have a sad future.
No, let us instead read silly poems!)

—FRANK WIKSTRÖM (TRANSLATOR)

Euler found a fascinating infinite product for pi, in which the numerators are the prime numbers larger than 2 and the denominators are even numbers that differ by one from the numerators and are not divisible by 4.

The primary purpose of the DATA statement is to give names to constants; instead of referring to pi as 3.141592653589793 at every appearance, the variable PI can be given that value with a DATA statement and used instead of the longer form of the constant. This also simplifies modifying the program, should the value of pi change. —FORTRAN manual for Xerox Computers

e to the u, du/dx
e to the x dx
cosine, secant,
tangent, sine,
3.14159
integral, radical,
u dv,
slipstick, slide rule,
MIT!

—Massachusetts Institute of Technology football cheer

In October 1996, Fabrice Bellard calculated the 400 billionth hexadecimal digit of pi using an equation created by Simon Plouffe, Peter Borwein, and Jonathan Borwein that lets you calculate the nth digit of pi without calculating any previous digits:

$$\pi = \sum_{n=0}^{\infty} \left(\frac{4}{8n+1} - \frac{2}{8n+4} - \frac{1}{8n+5} - \frac{1}{8n+6} \right) \times \left(\frac{1}{16} \right)^n$$

Unfortunately, you cannot convert this into a decimal (base 10) number without knowing every digit that comes before it. Nonetheless, the ability to calculate the nth digit of a transcendental number like pi was thought to be impossible.

In the Star Trek episode "Wolf in the Fold," Spock foils the evil computer by telling it to "compute to the last digit the value of pi."

CIRCLE DIGITS: A Self-Referential Story by Michael Keith

The number of letters in each word represents a digit of pi. Punctuation marks other than periods represent zero; words of more than ten letters represent the obvious two digits. A digit stands for itself (this is only used once).

For a time I stood pondering on circle sizes. The large computer mainframe quietly processed all of its assembly code. Inside my entire hope lay for figuring out an elusive expansion. Value: pi. Decimals expected soon. I nervously entered a format procedure. The mainframe processed the request. Error. I, again entering it, carefully retyped. This iteration gave zero error printouts in all—success. Intently I waited. Soon, roused by thoughts within me, appeared narrative mnemonics relating digits to verbiage! The idea appeared to exist but only in abbreviated fashion—little phrases typically. Pressing on I then resolved, deciding firmly about a sum of decimals to use—likely around four hundred, presuming the computer code soon halted! Pondering these ideas, words appealed to me. But a problem of zeros did exist. Pondering more, solution subsequently appeared. Zero suggests a punctuation element. Very novel! My thoughts were culminated. No periods, I concluded. All residual marks of punctuation = zeros. First digit expansion answer then came before me. On examining some problems unhappily arose. That imbecilic bug! The printout I possessed showed four nine as foremost decimals. Manifestly troubling. Totally every number looked wrong. Repairing the bug took much effort. A pi mnemonic with letters truly seemed good. Counting of all the letters probably should suffice. Reaching for a record would be helpful. Consequently, I continued, expecting a good final answer from computer. First number slowly displayed on the flat screen—3. Good. Trailing digits apparently were right also. Now my memory scheme must probably be implementable. The technique was chosen, elegant in scheme: by self reference a tale mnemonically helpful was ensured. An able title suddenly existed—"Circle Digits." Taking pen I began. Words emanated uneasily. I desired more synonyms. Speedily I found my (alongside me) Thesaurus. Rogets is probably an essential in doing this, instantly I decided. I wrote and erased more. The Rogets clearly assisted immensely. My story proceeded (how lovely!) faultlessly. The end, above all, would soon joyfully overtake. So, this memory helper story is incontestably complete. Soon I will locate publisher. There a narrative will I trust immediately appear, producing fame.

The end.

A Conversation Between a Human and an Alien Being in Carl Sagan's 1985 Novel Contact

"Your mathematicians have made an effort to calculate it out to . . . let's say the ten-billionth place. You won't be surprised to hear that other mathematicians have gone further. Well, eventually—let's say it's in the ten-to-the-twentieth place—something happens. The randomly varying digits disappear, and for an unbelievably long time there's nothing but ones and zeros. . . ."

"And the number of zeros and ones? Is it a product of prime numbers?"

"Yes, eleven of them."

"You're telling me that there's a message in eleven dimensions hidden deep inside the number pi? Someone in the universe communicates by . . . mathematics? . . . How can you hide a message inside pi? It's built into the fabric of the universe."

"Exactly."

She stared at him.

"It's even better than that," he continued. "Let's assume that only in base-ten arithmetic does the sequence of zeros and ones show up, although you'd recognize that something funny's going on in any other arithmetic. Let's also assume that the beings who first made this discovery had ten fingers. You see how it looks? It's as if pi has been waiting for billions of years for ten-fingered mathematicians with fast computers to come along. You see the Message was kind of addressed to us."

AFTERWORD

The magic of pi is not confined to the circle or to the measure of arcs and curves. While at first it appears that a circle defines pi, perhaps pi defines the circle instead. Certainly, even without knowing pi, we can create circles. And yet pi exists separate from circles as well. It resonates in us and around us. It is the answer to a thousand math puzzles, each without any relation to circles; it's within the solutions of probability and statistical questions; and it's part of how we interpret natural phenomena as varied as the structure of atoms to the motion of the stars.

The apparent simplicity of the circle and the square beckons us to define them when, in fact, they open door after door to the infinite mysteries that weave the fabric of our universe.

And yet, however hard we search, pi is ultimately unknowable. There is talk of harnessing the power of thousands of computers, connected via the Internet, to calculate digits of pi at night when the computers are least needed, until we can explore trillions of digits of the elusive number. But no matter how many computers we use or how much time we spend, we remain bound to the manifest and the measureable. Mathematician and pi researcher Peter Borwein has predicted that we will never know the $10^{51\text{st}}$ digit of pi; after all, there are probably only about 10^{78} atoms in the entire universe. But the digits of pi extend even beyond this number, its irregular pattern of digits reeling off far beyond our comprehension.

When you look into a mirror, you see the roundness of your eye—the colored iris forming a circular window through which, it is said, you can see your soul. Perhaps it is this roundness itself that first led the early humans to contemplate circles and, later, their measurement. Perhaps it is this roundness that will draw us further into the search for pi.

For More Information on Pi . . .

If this book has inspired you to search out even more pi-related information, you'll find most everything available on the subject among three sources. First, the Internet and the World Wide Web offer a great deal in the way of both serious and frivolous discussions on pi, and the place to start your Web-surfing exploration is: http://www.joyofpi.com.

If you favor paper-based information, you might try Petr Beckmann's A History of Pi (1971, St. Martin's Press) or Pi: A Source Book, by Len Berggren, Jonathan Borwein, and Peter Borwein (1997, Springer-Verlag). Know that Beckmann's book is a quirky stroll through pi's history; the book by Berggren, et al., is a 750-page reference that includes every major piece written about pi in the past three millennia.

INDEX

A

B

C

D

E

F

G

H

I

J

K

L

n

o

p

Q

R

S

T

U

V

W

Z

1

1 MILLIONTH DIGIT